Contents

WITHDRAWN
FOR
REFERENCE ONLY

How this book will help you

by John Berry, Ted Graham and Roger Williamson

Exam practice – how to answer questions better

This book will help you to improve your performance in your AS Maths exam.

The key to success in Maths is lots of practice. In this book we have provided numerous **AS exam questions for you to try to answer, along with support to help you** if you get stuck. The two most common reasons that students lose marks in exams are that they make silly algebraic errors or that they are not familiar with the topics on which the questions are based. We have provided **good coverage of the main AS Maths topics** (Core Maths, Mechanics, Statistics and Decision Maths) so that you can make sure you are familiar with the key topics. We have included two chapters on algebra to help you practise and revise your algebraic skills.

Each chapter in this book is broken down into four separate elements, aimed at giving you as much practice and guidance as possible:

1 'Key points to remember' and 'Don't make these mistakes'

On these pages we outline the key points that you need to know for each of the topics. These points should not be new to you and our aim is to provide a **reminder that will help refresh your memory.** We also include **lists of formulae that you must not forget.** In the new AS examination, you are expected to learn formulae rather than be able to look them up, so this section is very important. We also include **checklists of common mistakes that you should try to avoid.**

2 Exam Questions and Answers with 'How to score full marks'

We have used a number of exam-style questions to illustrate **the methods that you will need to use in your exam.** We provide a question which is followed by a sample correct solution. Alongside this **we provide notes to guide you through the solution and indicate the steps needed to score full marks.**

3 Questions to try

This is where **you get the opportunity to practise answering the types of question that you will meet in your exam.** Each section starts with some easier questions to warm you up and then includes actual exam questions or exam-style questions. **You need to attempt all the questions as this sort of practice will help you to reach your full potential in your exam.**

4 Answers to the Questions to try

At the back of the book you will find **solutions to all of the Questions to try**. Try answering the questions as though you were in the exam. Look at the answers if you are really stuck or when you think that you have completed the question correctly. **Alongside each solution we have written a commentary that identifies the key stages in the solution – this will help you if you get stuck.** The commentary will also help you to make sure that you have not omitted any important working that you should have shown.

This book includes questions on Core Maths, Mechanics, Statistics and Decision Maths. It focuses on the most important or difficult parts of the AS Mathematics core, which all exam boards must cover for an AS award. Some of the exam boards have chosen to go beyond this core and you will find material for these topics in our A2 Mathematics book. The table below shows how the topics in this book fit into the modules for the different exam boards. Some sections may contain a little more material than is required for your particular exam.

The topics covered by your specification

Chapters in this book	AQA	EDEXCEL	OCR	OCR (MEI)
1. Algebra and equations	C1	C1	C1	C1
2. Arithmetic and geometric progressions	C2	C1/2	C2	C2
3. Trigonometry	C2	C2	C2	C2
4. Coordinate geometry	C1	C1/2	C1	C1
5. The binomial expansion	C2	C2	C2	C1
6. Differentiation	C1/2	C1/2	C1	C2
7. Integration	C1/2	C1/2	C2	C2
8. The trapezium rule	C2	C2	C2	C2
9. Factor and remainder theorems	C1	C2	C2	C1
10. Exponentials and logarithms	C2	C2	C2	C2
11. Transformations	C2	C1	C1	C1
12. Kinematics on a straight line	M1	M1	M1	M1
13. Kinematics and vectors	M1	M1		M1
14. Newton's laws and connected particles	M1	M1	M1	M1
15. Conservation of momentum	M1	M1	M1	M1
16. Projectiles	M1	M2	M2	M1
17. Numerical measures	S1	S1	S1	S1
18. Probability	S1	S1	S1	S1
19. Binomial distribution	S1	S2	S1	S1
20. Normal distribution	S1	S1	S2	S2
21. Correlation and regression	S1	S1	S1	S2
22. Algorithms	D1	D1	D1	D1
23. Networks	D1	D1	D1	D1
24. Critical path analysis	D2	D1	D1	D1
25. Optimisation	D1	D1	D2	D1

- There are two types of marks available to you: **method marks and accuracy marks**. Accuracy marks are awarded for correct working and answers. Method marks are awarded for working that is of the type required, but where there is a reasonably minor error.

  For example, if when solving a quadratic equation you fail to include a negative sign with one of the numbers, you would lose an accuracy mark, but may be awarded a method mark because you have shown that you know how to solve a quadratic equation. **In order to gain method marks, it is very important that you show your working clearly so that the examiner can give you credit if the answer is wrong.** Short comments to explain to the examiner what you are trying to do may help.

- **In 'show that...' questions, make sure that you really do show that the result is true.** You should include all the necessary steps in your working. Students often miss out obvious – but important – steps when answering these types of question.

- **Make sure that you have learned the formulae that you need to know for each topic.** You may like to make a list that you can spend time studying. Imagine sitting in the exam looking at a question, but being unable to remember the formula that you need. Also be aware of which formulae are in the formulae book for your exam board and be able to find them when you need them.

- **Be aware of which papers for your board have calculator restrictions.** This will apply to one of the Core Maths papers. You should know which paper does not allow the use of calculators. Graphic calculators can be used in all of the other papers. A graphics calculator can be an advantage in an exam, but you do need to have spent time getting to know how to use it.

- When you are asked for an exact answer or are asked to 'show that $x = \dfrac{\sqrt{3}}{7}$', or similar, **do not use decimal approximations as you are working.** Always work with surds or fractions until you obtain the required result.

- **Look at the number of marks that are awarded for each part of a question.** This will give you a guide to how much work you have to do. A question that has one mark for an explanation or comment will expect one sentence. Do not write half a page of explanation, as you will be wasting time.

- **Use the time that you have available in the exam productively.** If you know that you are not progressing with a question, stop and move on to another one. Also, always try the later parts of questions, even if you cannot do the first part. Exam questions often use the 'show that...' format so that students can attempt later parts even if they can't do the 'show that...' part.

1 Algebra and equations

Key points to remember

Techniques

- A surd is a number of the form $a + \sqrt{b}$ where a and b are integers.
- Use the difference of two squares to remove a surd in the denominator of an expression, e.g. for $a + \sqrt{b}$, multiply both denominator and numerator by $a - \sqrt{b}$ since $(a + \sqrt{b})(a - \sqrt{b}) = a^2 - b$.

Quadratic equations

- Any quadratic function $ax^2 + bx + c$ can be written in the form $a(x + p)^2 + q$. This is called **completing the square**.
- A quadratic equation of the form $ax^2 + bx + c = 0$ can be solved using the **formula**:
$$x = \frac{-b \pm \sqrt{b^2 - 4ac}}{2a}$$
- The equation $ax^2 + bx + c = 0$ has real roots if $b^2 - 4ac \geqslant 0$.

Formulae you must know

- $x^a x^b = x^{a + b}$
- $x^a \div x^b = x^{a - b}$
- $(x^a)^b = x^{a \times b} = x^{ab}$
- $x^0 = 1$
- $x^{-a} = \dfrac{1}{x^a}$
- Solutions of $ax^2 + bx + c = 0$ are:
$$x = \frac{-b \pm \sqrt{b^2 - 4ac}}{2a}$$
- $(a + \sqrt{b})(a - \sqrt{b}) = a^2 - b$
- $(a + b)(a - b) = a^2 - b^2$

Graphs of quadratic functions

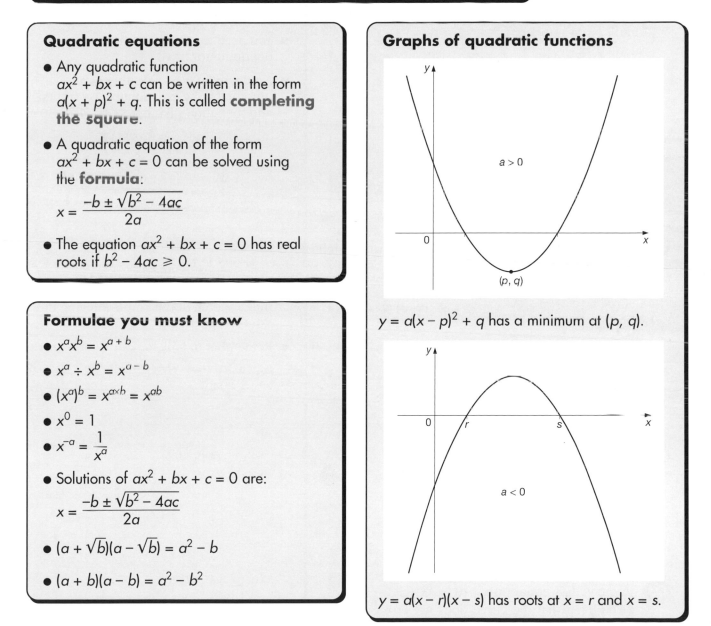

$y = a(x - p)^2 + q$ has a minimum at (p, q).

$y = a(x - r)(x - s)$ has roots at $x = r$ and $x = s$.

Exam Questions and Student's Answers

Q1 Express $\dfrac{(\sqrt{3} - 1)^2}{\sqrt{3} + 1}$ in the form $a + b\sqrt{3}$, where a and b are integers.

$$\dfrac{(\sqrt{3} - 1)^3}{(\sqrt{3} + 1)(\sqrt{3} - 1)} = \dfrac{(\sqrt{3} - 1)^3}{3 - 1}$$

$$= \dfrac{(\sqrt{3})^3 - 3(\sqrt{3})^2 + 3\sqrt{3} - 1}{2}$$

$$= \dfrac{6\sqrt{3} - 10}{2}$$

$$= -5 + 3\sqrt{3}$$

Q2 Find the range of values for x for which $(x - 4)(2x + 3) < 0$.

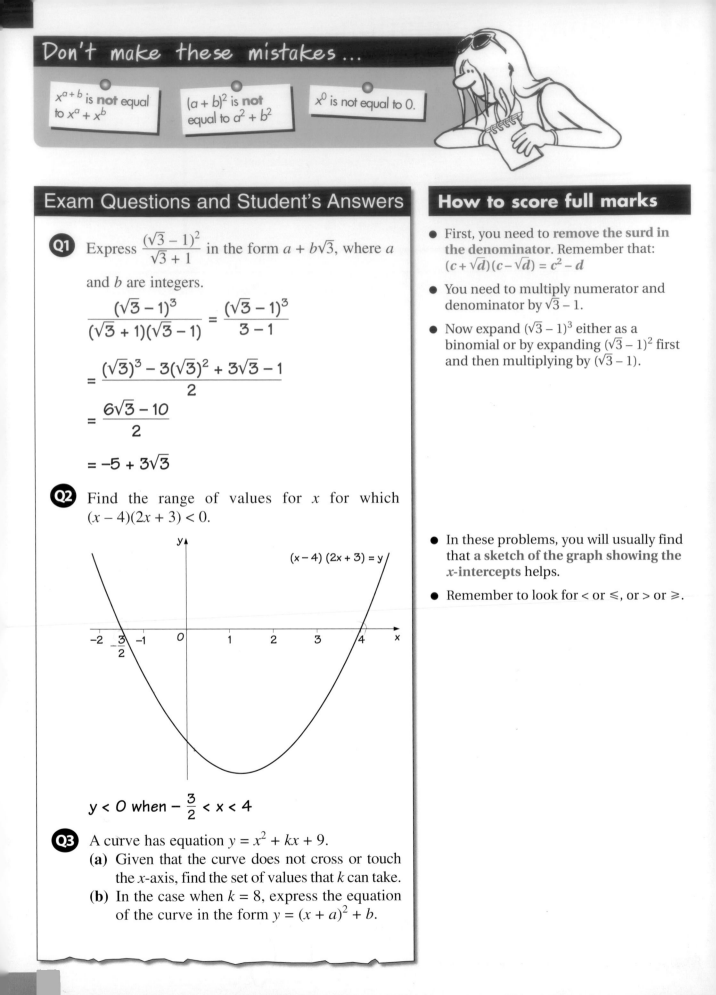

$(x-4)(2x+3) = y$

$y < 0$ when $-\dfrac{3}{2} < x < 4$

Q3 A curve has equation $y = x^2 + kx + 9$.
   (a) Given that the curve does not cross or touch the x-axis, find the set of values that k can take.
   (b) In the case when $k = 8$, express the equation of the curve in the form $y = (x + a)^2 + b$.

How to score full marks

- First, you need to **remove the surd in the denominator**. Remember that:
  $(c + \sqrt{d})(c - \sqrt{d}) = c^2 - d$

- You need to multiply numerator and denominator by $\sqrt{3} - 1$.

- Now expand $(\sqrt{3} - 1)^3$ either as a binomial or by expanding $(\sqrt{3} - 1)^2$ first and then multiplying by $(\sqrt{3} - 1)$.

- In these problems, you will usually find that **a sketch of the graph showing the x-intercepts** helps.

- Remember to look for $<$ or $\leqslant$, or $>$ or $\geqslant$.

(c) Find the solutions of the equation
$x^2 + kx + 9 = 0$ for x, given that $k > 6$.

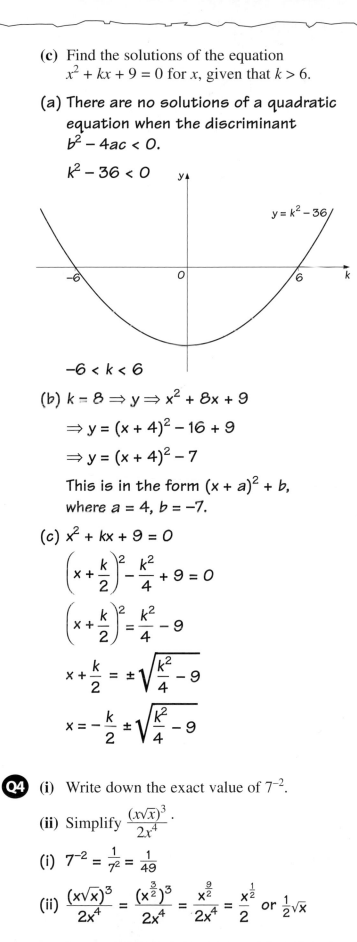

(a) There are no solutions of a quadratic equation when the discriminant $b^2 - 4ac < 0$.

$k^2 - 36 < 0$

$-6 < k < 6$

(b) $k = 8 \Rightarrow y \Rightarrow x^2 + 8x + 9$

$\Rightarrow y = (x + 4)^2 - 16 + 9$

$\Rightarrow y = (x + 4)^2 - 7$

This is in the form $(x + a)^2 + b$, where $a = 4$, $b = -7$.

(c) $x^2 + kx + 9 = 0$

$\left(x + \dfrac{k}{2}\right)^2 - \dfrac{k^2}{4} + 9 = 0$

$\left(x + \dfrac{k}{2}\right)^2 = \dfrac{k^2}{4} - 9$

$x + \dfrac{k}{2} = \pm\sqrt{\dfrac{k^2}{4} - 9}$

$x = -\dfrac{k}{2} \pm\sqrt{\dfrac{k^2}{4} - 9}$

Q4 **(i)** Write down the exact value of 7^{-2}.

(ii) Simplify $\dfrac{(x\sqrt{x})^3}{2x^4}$.

(i) $7^{-2} = \dfrac{1}{7^2} = \dfrac{1}{49}$

(ii) $\dfrac{(x\sqrt{x})^3}{2x^4} = \dfrac{(x^{\frac{3}{2}})^3}{2x^4} = \dfrac{x^{\frac{9}{2}}}{2x^4} = \dfrac{x^{\frac{1}{2}}}{2}$ or $\dfrac{1}{2}\sqrt{x}$

How to score full marks

- Remember that **a positive number has two square roots**, so the solution is not simply $k < 6$.

- A sketch graph of $y = k^2 - 36$ helps.

- **Substitute $k = 8$ into the equation** and complete the square.

- This problem requires you to **complete the square**. In this case you need to use $\frac{k}{2}$ as the second term in the bracket.

- Once you have completed the square the equation can be solved using the steps shown.

- Don't forget to **introduce a ± sign when you square root both sides** of the equation.

- You need an exact answer, so don't use your calculator.

- Simplify or expand the numerator first.

Q1 Simplify each of the following expressions.

(a) $a^5 \times a^{-2}$

(b) $\dfrac{a^5}{a^{-3}}$

(c) $\left(\dfrac{\sqrt[4]{a}}{\sqrt[7]{a}} \right)^3$

Q2 A quadratic function is defined by:

$$f(x) = x^2 + kx + 9$$

where k is a constant. It is given that the equation $f(x) = 0$ has two distinct real roots. Find the set of values that k can take.

For the case where $k = -4\sqrt{3}$:

(i) express $f(x)$ in the form $(x + a)^2 + b$, stating the values of a and b, and hence write down the least value taken by $f(x)$

(ii) solve the equation $f(x) = 0$, expressing your answer in terms of surds, simplified as far as possible.

Q3 Solve the inequality $x^2 > x + 20$.

Q4 Two consecutive integers are chosen so that their sum is greater than 10 and their product is less than 72. Find the range of values within which the two numbers must be.

Q5 The specification for a new rectangular car park states that its length, x m, is to be 5 m more than its width. The perimeter of the car park is to be greater than 32 m.
(a) Form and solve a linear inequality in x.
(b) The area of the car park is to be less than 104 m². Form a quadratic inequality in x.
(c) Solve your quadratic inequality and state the possible range of values of the length of the car park.

Q6 (i) Solve the simultaneous equations.

$$y = x^2 - 3x + 2 \qquad y = 3x - 7$$

(ii) Interpret your solution to part (i) geometrically.

Q7 The quadratic equation $x^2 + kx + k = 0$ has no real roots for x.
(a) Write down the discriminant of $x^2 + kx + k = 0$ in terms of k.
(b) Hence find the set of values that k can take.

Q8 The number x satisfies the equation $x^2 + mx + 16 = 0$, where m is a constant.
Find the values of m for which the equation has:
(a) equal roots
(b) two distinct roots
(c) no real roots.

Q9 (a) Express $x^2 - 8x + 3$ in the form $(x + a)^2 + b$.
(b) Hence write down the coordinates of the minimum point on the graph of $y = x^2 - 8x + 3$.

Q10 (a) Express $\dfrac{\sqrt{2} + 1}{\sqrt{2} - 1}$ in the form $a\sqrt{2} + b$, where a and b are integers.
(b) Solve the inequality $\sqrt{2}(x - \sqrt{2}) < x + 2\sqrt{2}$.

Answers can be found on pages 125–126.

2 Arithmetic and geometric progressions

Key points to remember

- **A sequence is a list of terms** defined by some relationship,
  e.g. 3, 7, 11, 15, 19, ...

- **A series is the sum of the terms of a sequence**,
  e.g. 3 + 7 + 11 + 15 + 19 + ...

- **A sequence may converge to a limit**,
  e.g. $1, \frac{1}{2}, \frac{1}{4}, \frac{1}{8}, ...$
  which **converges to 0**

  or it may **diverge**,
  e.g. 1, 4, 16, 64, ...
  which **increases to infinity**.

- **A series may converge**,
  e.g. $1 + \frac{1}{3} + \frac{1}{9} + \frac{1}{27} + ...$
  which **converges to $\frac{3}{2}$**

  or it may **diverge**,
  e.g. 1 + 3 + 5 + 7 + 9 + ...
  which **increases to infinity**.

- **Arithmetic progression**

  If the **first term** is a and the **common difference** is d, then the **nth term u_n** is given by: $u_n = a + (n - 1)d$ and the **sum of the first n terms S_n** is given by:

  $S_n = \frac{1}{2}n(2a + (n - 1)d)$ or $S_n = \frac{1}{2}n(a + l)$

  where l is the last term of the sequence. The relationship between successive terms is given by:

  $u_{n+1} = u_n + d$

- **Geometric progression**

  If the **first term** is a and the **common ratio** is r, then the **nth term u_n** is given by:

  $u_n = ar^{n-1}$
  and the **sum of the first n terms S_n** is given by:

  $S_n = a\left(\dfrac{r^n - 1}{r - 1}\right)$ or $S_n = a\left(\dfrac{1 - r^n}{1 - r}\right)$

  The relationship between successive terms is given by:
  $u_{n+1} = u_n \times r$

- **Sum to infinity of a GP**

  If the **common ratio r** of a GP is such that $-1 < r < 1$, then the sum to infinity of the GP is given by:

  $S_\infty = \dfrac{a}{1 - r}$

- **Formulae you must know how to use**

  These will be in your formulae book.

  - $u_n = a + (n - 1)d$
  - $S_n = \frac{1}{2}n(2a + (n - 1)d)$
  - $u_n = ar^{n-1}$
  - $S_n = a\left(\dfrac{1 - r^n}{1 - r}\right)$
  - $S_\infty = \dfrac{a}{1 - r}$

Don't make these mistakes...

Don't use the **wrong number of terms**.

Don't forget to **subtract 1 from n** when finding the **nth term**.

Don't use the **sum to infinity formula** without checking that $-1 < r < 1$.

Q1 A sequence has the terms: 2, 5, 8, 11, 14, ...
(a) Find the 100th term of the sequence.
(b) Find the sum of the first 100 terms of the sequence.

(a) $u_{100} = 2 + (100 - 1) \times 3$

$= 2 + 99 \times 3$

$= 299$

(b) $S_n = \frac{1}{2} \times 100 \times (2 \times 2 + (100 - 1) \times 3)$

$= 50 \times 301$

$= 15\,050$

- You need to **recognise that this is an arithmetic progression** because there is a constant difference between adjacent terms. The sequence has common difference 3 and first term 2, so $d = 3$ and $a = 2$.

- Use the formula $u_n = a + (n - 1)d$, with $n = 100$.

- Use the formula $S_n = \frac{1}{2}n(2a + (n - 1)d)$, with $n = 100$.

Q2 A geometric progression, with positive terms, has first term 4 and third term 2.56.
(a) Find the sum of the first five terms.
(b) Find the sum to infinity of the terms of the progression.

(a) $4r^2 = 2.56$

$r = \sqrt{\dfrac{2.56}{4}} = 0.8$

$S_5 = 4\left(\dfrac{1 - 0.8^5}{1 - 0.8}\right) = 13.4$

correct to three significant figures.

(b) $S_\infty = \dfrac{4}{1 - 0.8} = 20$

- Your first task is to **find the common ratio** for the GP. To do this, substitute $a = 4$, $n = 3$ and $u_3 = 2.56$ into the formula $u_n = ar^{n-1}$ and solve for r.

- Then **use the formula** $S_n = a\left(\dfrac{1 - r^n}{1 - r}\right)$ to find the sum of the terms of the GP.

- Finally, **use the formula** $S_\infty = \dfrac{a}{1 - r}$ to find the sum to infinity.

Q3 At the start of each year a person pays £200 into a savings account. Interest is added at a rate of 8% per year at the end of each year. Find the value of the investment after:
(a) 2 years
(b) 20 years.

(a) The value of the investment after 1 year is given by:

$200 \times 1.08 = £216$

and at the end of year 2 the value is:
$200 \times 1.08^2 + 200 \times 1.08 = £449.28.$

- For this type of problem it's a good idea to look at some simple cases, where there are just a few terms, to help **establish how a sequence or series can be built up**. In this question you can do this in part (a).

(b) The value of the investment after 20 years is given by:

$$200 \times 1.08 + 200 \times 1.08^2 + 200 \times 1.08^3 + \ldots + 200 \times 1.08^{19} + 200 \times 1.08^{20}$$

GP with $r = 1.08$, $a = 200 \times 1.08 = 216$ and $n = 20$, which gives:

$$S_{20} = 216 \times \left(\frac{1.08^{20} - 1}{1.08 - 1} \right)$$

$$= £9884.58$$

Q4 The 8th term of an AP is 45 and the sum of the first 12 terms is 468.
(a) Find the common difference and the first term.
(b) If the sum of the first r terms is 752, find r.

(a) $u_n = a + (n - 1)d$

So $45 = a + 7d$ (1)

$$S_n = \tfrac{1}{2}n(2a + (n - 1)d)$$

So $468 = 6(2a + 11d)$

$78 = 2a + 11d$ (2)

$90 = 2a + 14d$ $2 \times$ (1)

$78 = 2a + 11d$ (2)

Subtracting gives:

$12 = 3d$

$d = 4$

Substituting the value of d back into equation (1):

$45 = a + 28$

so that $a = 17$.

(b) $S_n = \tfrac{1}{2}n(2a + (n - 1)d)$

So $752 = \tfrac{1}{2}r(2 \times 17 + (r - 1) \times 4)$

$0 = 2r^2 + 15r - 752$

$$r = \frac{-15 \pm \sqrt{15^2 - 4 \times 2 \times (-752)}}{2 \times 2}$$

$= 16$ or -23.5

$\therefore r = 16$

How to score full marks

● In part (b) it is important to **see how the terms of the progression develop and to identify the values of r, a and n that are needed.**

● Once you have these, you can substitute them into the formula
$$S_n = a\left(\frac{r^n - 1}{r - 1} \right)$$

● First **use the formula for the nth term** to form one equation.

● Then **use the formula for the sum of n terms** to form a second equation, giving a pair of **simultaneous equations.**

● **Solve these** to find a value for d.

● Use this value of d to find a.

● Use the formula for the sum of n terms with $a = 17$, $d = 4$ and $S_r = 752$, to form a **quadratic equation.**

● **Solve the quadratic equation** to find the possible values or r.

● As r must be a **natural number,** disregard the value of -23.5.

Questions to try

Q1 Find the 18th term and the sum of the first 15 terms of the sequence that begins 7, 9, 11, 13, 15, …

Q2 A sequence begins: 100, 90, 81, 72.9, …
(a) Find the 10th term of the sequence, correct to three decimal places.
(b) Find the sum of the first 20 terms of the sequence.
(c) Find the sum to infinity of the sequence.

Q3 The first term of an arithmetic series is 4 and the common difference 5.
(a) Calculate the sum of the first 20 terms.
(b) Find the smallest value of n such that the nth term of the series is greater than 131.

Q4 The 8th term of an AP is 40 and the 20th term is 124. Find the first term and the common difference. Also find the sum of the first 20 terms.

Q5 The second term of a geometric series is 80 and the fifth term of the series is 5.12.
(a) Find the common ratio and the first term of the series.
(b) Find the sum to infinity of the series, giving your answer as an exact fraction.
(c) Find the difference between the sum to infinity of the series and the sum of the first 14 terms of the series, giving your answer in the form $a \times 10^n$, where $1 \leqslant a < 10$ and n is an integer.

Q6 The sum to infinity of a GP is exactly 7.2 and the second term of the sequence is 1. Find the two possible common ratios. For the larger of these, find the first four terms of the sequence and the sum of the first 20 terms correct to three decimal places.

Q7 The amount paid into a pension fund increases by $r\%$ each year. In the first year £1200 is paid in and in the fifth year £1932 is paid into the fund. Find r and the total paid into the fund over 10 years.

Q8 A company wants to recruit people to work for a 10-month period. They offer two different types of pay scheme.
A   Starting salary of £1000 per month, increasing by £100 per month
B   Starting salary of £X, increasing by 10% per month
Find X so that the total paid by both schemes is the same.

Q9 A ball rebounds to $\frac{2}{5}$ of the height from which it was dropped. The ball is dropped from a height of 2 m. Find the distance travelled by the ball, when it hits the ground for the second time. Find the total distance travelled by the ball, before it stops.

Q10 Records are kept of the number of copies of a certain book that are sold each week. In the first week after publication 3000 copies were sold, and in the second week 2400 copies were sold. The publisher forecasts future sales by assuming that the number of copies sold each week will form a geometric progression with first two terms 3000 and 2400. Calculate the publisher's forecasts for:
(i) the number of copies that will be sold in the 20th week after publication
(ii) the total number of copies sold during the first 20 weeks after publication
(iii) the total number of copies that will ever be sold.

Answers can be found on pages 126–127.

3 Trigonometry

Key points to remember

- **Convert to and from radians** and degrees using:
  $180° = \pi$ radians
- **Arc length** $= r\theta$ when θ is in radians.
- **Area of a sector** $= \frac{1}{2}r^2\theta$ when θ is in radians.
- Remember that **trigonometric equations may have more than one solution.**
- **Know how to find the solutions in a given range.** For example the equation $\sin 2x = 0.5$ has four solutions in the interval $0 \leqslant x \leqslant 360°$, as shown on the graph.

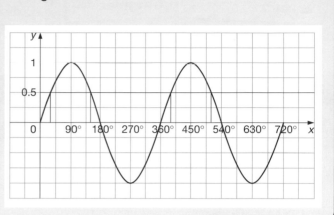

- The graph below shows $y = A\sin(\omega x + \alpha) + d$.

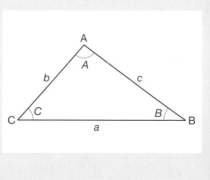

NO, JAMIE, NOT TWO PIGS WITH RAISINS – I SAID THERE ARE TWO PI RADIANS IN THE MIDDLE!

- **Pythagorean identity**
  $\cos^2\theta + \sin^2\theta = 1$
- **Sine rule**
  $$\frac{a}{\sin A} = \frac{b}{\sin B} = \frac{c}{\sin C}$$
- **Cosine rule**
  $a^2 = b^2 + c^2 - 2bc\cos A$
- Know that:
  $\tan\theta = \dfrac{\sin\theta}{\cos\theta}$

Formulae you must know

- Arc length $= r\theta$
- Area of a sector $= \frac{1}{2}r^2\theta$
- $\cos^2\theta + \sin^2\theta = 1$
- $\tan\theta = \dfrac{\sin\theta}{\cos\theta}$
- $\dfrac{a}{\sin A} = \dfrac{b}{\sin B} = \dfrac{c}{\sin C}$
- $a^2 = b^2 + c^2 - 2bc\cos A$
- $A = \frac{1}{2}ab\sin C$

Don't make these mistakes...

Don't forget to include all the solutions when solving an equation.

Don't mix up radians and degrees.

Don't make arithmetic or algebraic errors when rearranging equations.

Exam Questions and Student's Answers

How to score full marks

Q1 The diagram shows a circle of radius 4 cm and centre O. The two points, A and B, lie on the circle such that $\angle AOB = \dfrac{\pi}{6}$ radians.

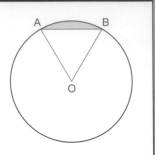

Find the area of the shaded region.

Area of sector $= \dfrac{1}{2} \times 4^2 \times \dfrac{\pi}{6}$

$= \dfrac{4}{3}\pi \text{ cm}^2$

Area of triangle $= \dfrac{1}{2} \times 4 \times 4\sin\dfrac{\pi}{6}$

$= 4 \text{ cm}^2$

Area of shaded region

$= \dfrac{4}{3}\pi - 4$

$= 0.189 \text{ cm}^2$ to three decimal places.

- **First find the area of the sector** using the formula $A = \dfrac{1}{2}r^2\theta$.

- **To find the area of the triangle, draw in the height of the triangle**, form an expression for the height and then calculate the area of the triangle.

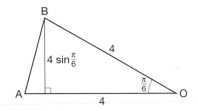

- **The area of the shaded region** is the area of the sector *minus* the area of the triangle.

Q2 Solve the equation $4\sin x = 3$, giving all the solutions in the range $0 \leqslant x \leqslant 360°$.

$4\sin x = 3$

$\sin x = \dfrac{3}{4}$

$x = 48.6°$ or $x = 180° - 48.6° = 131.4°$

$x = 48.6°$ or $x = 131.4°$

- **First rewrite the equation in the form** $\sin x = \dots$.

- **The graph shows that there are two solutions in the range $0 \leqslant x \leqslant 360°$.**

- If you use a calculator you find $x = 48.6°$ (correct to one decimal place) as the first solution. From the graph you can see that this value must be subtracted from 180° to get the second solution.

Q3 Solve the equation $4\cos 2x + 1 = 3$, giving all the solutions in the range $0 \leqslant x \leqslant 360°$.

$4 \cos 2x + 1 = 3$

So $\cos 2x = \frac{1}{2}$

This gives $2x = 60°$. Other values of $2x$ that satisfy this equation can be found from the graph of $y = \cos x$.

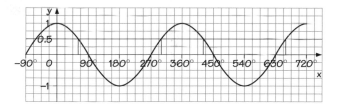

The other solutions are:

$2x = 360° - 60° = 300°,$
$2x - 360° + 60° = 420°,$
$2x = 720° - 60° = 660°$

$60° \div 2 = 30°, 300° \div 2 = 150°,$
$420° \div 2 = 210°, 660° \div 2 = 330°$
$x = 30°, 150°, 210°, 330°$

Q4 Solve the equation $5 = 6\cos^2\theta + \sin\theta$, giving all the solutions in the range $0 \leqslant \theta \leqslant 360°$.

$5 = 6\cos^2\theta + \sin\theta$
$5 = 6(1 - \sin^2\theta) + \sin\theta$
$6\sin^2\theta - \sin\theta - 1 = 0$
$(2\sin\theta - 1)(3\sin\theta + 1) = 0$
$\sin\theta = \frac{1}{2}$ and $\sin\theta = -\frac{1}{3}$

For $\sin\theta = \frac{1}{2}$, the solutions are $\theta = 30°$

and $\theta = 180° - 30° = 150°$.

For $\sin\theta = -\frac{1}{3}$, a calculator will give $\theta = -19.5°$. The solutions in the range $0 \leqslant \theta \leqslant 360°$ are
$\theta = 180° + 19.5° = 199.5°$ and
$\theta = 360° - 19.5° = 340.5°$.

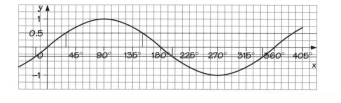

- Once you have written the equation in the form $\cos 2x = \ldots$ then you can find a value for $2x$. **You should recognise that $2x = 60°$ in this case**, but at other times you may need to use a calculator.

- As the equation contains $2x$, you first need to **find all the solutions between 0 and 720°**, so that **when they are halved you obtain solutions in the range 0 to 360°.**

- The graph shows the solution for $2x$.

- These are the possible values of $2x$. The corresponding values of x can now be found by dividing by 2.

- **Using the identity $\cos^2\theta + \sin^2\theta = 1$**, in the form $\cos^2\theta = 1 - \sin^2\theta$, you can eliminate the $\cos^2\theta$ term from the equation, to produce a **quadratic equation.**

- In this problem it is easy to **factorise the equation**, but in other cases you may need to use the **quadratic equation formula.**

- As there are **two solutions** to the quadratic equation there are **four values for θ**, as shown on the graph.

- Note that when you use $\sin^{-1}$ on your calculator for negative values, you will get a negative answer. **You need to be able to calculate angles in the range 0 to 360°.**

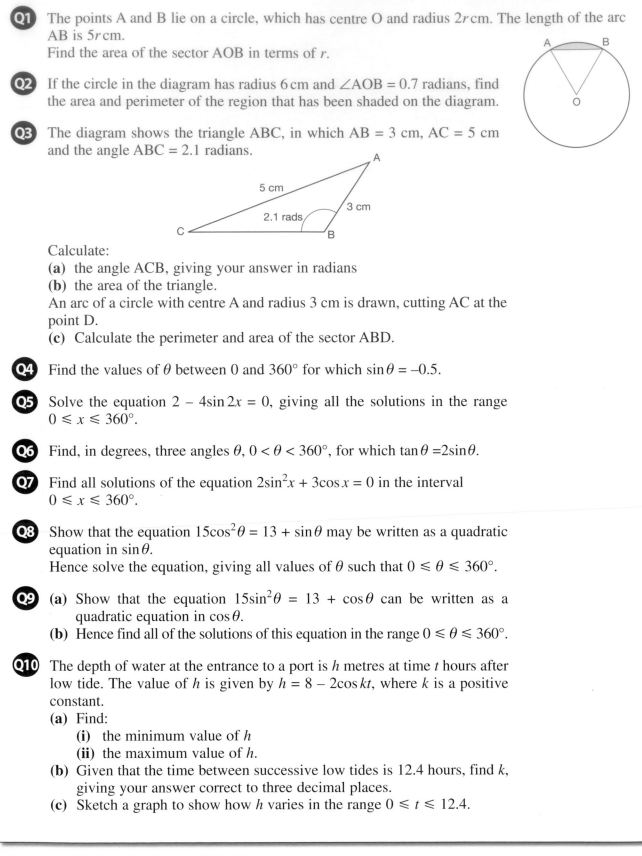

Q1 The points A and B lie on a circle, which has centre O and radius $2r$ cm. The length of the arc AB is $5r$ cm.
Find the area of the sector AOB in terms of r.

Q2 If the circle in the diagram has radius 6 cm and $\angle AOB = 0.7$ radians, find the area and perimeter of the region that has been shaded on the diagram.

Q3 The diagram shows the triangle ABC, in which AB = 3 cm, AC = 5 cm and the angle ABC = 2.1 radians.

Calculate:
(a) the angle ACB, giving your answer in radians
(b) the area of the triangle.
An arc of a circle with centre A and radius 3 cm is drawn, cutting AC at the point D.
(c) Calculate the perimeter and area of the sector ABD.

Q4 Find the values of θ between 0 and 360° for which $\sin\theta = -0.5$.

Q5 Solve the equation $2 - 4\sin 2x = 0$, giving all the solutions in the range $0 \leqslant x \leqslant 360°$.

Q6 Find, in degrees, three angles θ, $0 < \theta < 360°$, for which $\tan\theta = 2\sin\theta$.

Q7 Find all solutions of the equation $2\sin^2 x + 3\cos x = 0$ in the interval $0 \leqslant x \leqslant 360°$.

Q8 Show that the equation $15\cos^2\theta = 13 + \sin\theta$ may be written as a quadratic equation in $\sin\theta$.
Hence solve the equation, giving all values of θ such that $0 \leqslant \theta \leqslant 360°$.

Q9 **(a)** Show that the equation $15\sin^2\theta = 13 + \cos\theta$ can be written as a quadratic equation in $\cos\theta$.
(b) Hence find all of the solutions of this equation in the range $0 \leqslant \theta \leqslant 360°$.

Q10 The depth of water at the entrance to a port is h metres at time t hours after low tide. The value of h is given by $h = 8 - 2\cos kt$, where k is a positive constant.
(a) Find:
  (i) the minimum value of h
  (ii) the maximum value of h.
(b) Given that the time between successive low tides is 12.4 hours, find k, giving your answer correct to three decimal places.
(c) Sketch a graph to show how h varies in the range $0 \leqslant t \leqslant 12.4$.

Answers can be found on pages 127–128.

4 Coordinate geometry

Key points to remember

- The **equation of a straight line** can be written as $y = mx + c$, where m is the gradient and c is the intercept with the vertical axis.
- The **straight line through two points** with coordinates (x_1, y_1) and (x_2, y_2) has gradient $\dfrac{y_2 - y_1}{x_2 - x_1}$ and equation $\dfrac{y - y_1}{y_2 - y_1} = \dfrac{x - x_1}{x_2 - x_1}$.
- If the **gradient of a line is** m, then the **gradient of a line perpendicular to it is** $-\dfrac{1}{m}$.
- The **product of the gradients of two perpendicular lines** is -1.
- The **distance between the points** with coordinates (x_1, y_1) and (x_2, y_2) is $\sqrt{(x_2 - x_1)^2 + (y_2 - y_1)^2}$.
- The **midpoint of the line joining two points** with coordinates (x_1, y_1) and (x_2, y_2) is $\left(\dfrac{x_1 + x_2}{2}, \dfrac{y_1 + y_2}{2}\right)$.
- The **equation of a circle** with centre at (a, b) and with radius r is $(x - a)^2 + (y - b)^2 = r^2$. Note that for a circle with its centre at the origin this becomes $x^2 + y^2 = r^2$.

Formulae you must know

- $y = mx + c$
- $m = \dfrac{y_2 - y_1}{x_2 - x_1}$
- $\dfrac{y - y_1}{y_2 - y_1} = \dfrac{x - x_1}{x_2 - x_1}$
- $\sqrt{(x_2 - x_1)^2 + (y_2 - y_1)^2}$
- $\left(\dfrac{x_1 + x_2}{2}, \dfrac{y_1 + y_2}{2}\right)$
- $(x - a)^2 + (y - b)^2 = r^2$

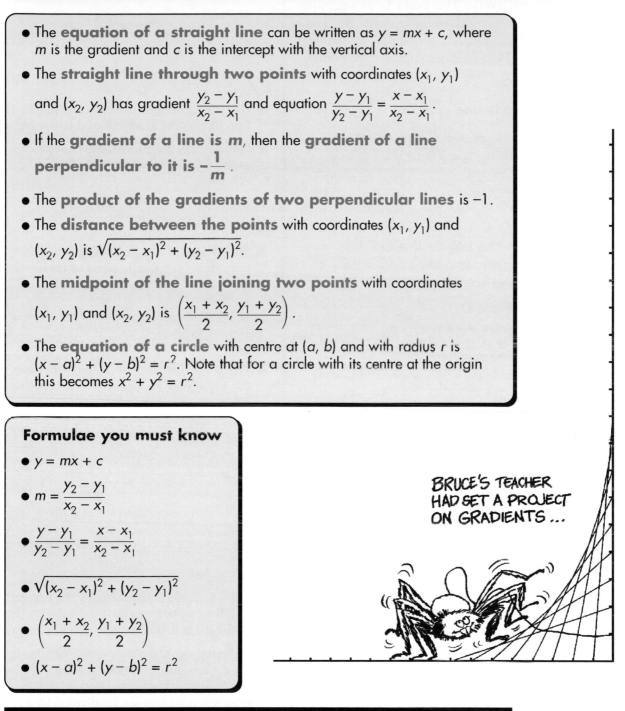

BRUCE'S TEACHER HAD SET A PROJECT ON GRADIENTS ...

Don't make these mistakes ...

Don't omit the negative sign when finding the gradient of a perpendicular to a line.

Don't make silly mistakes when using **negative coordinates** to find gradients.

Don't forget to **calculate the y-coordinate** when you have found the x-coordinate.

Q1 The points A and B have coordinates (7, 1) and (−1, −3) respectively. The point C is such that ABC is a right-angled triangle, with the right angle at C. The side AC is parallel to the line $3y + x - 12 = 0$. Find the coordinates of the point C.

$3y + x - 12 = 0$

$y = -\frac{1}{3}x + 4$ so it has gradient $-\frac{1}{3}$.

The line through A and C has equation $y = -\frac{1}{3}x + c$. Using $x = 7$ and $y = 1$ gives:

$1 = -\frac{7}{3} + c \Rightarrow c = \frac{10}{3}$

$y = -\frac{1}{3}x + \frac{10}{3}$

A line through the points B and C has gradient 3. Its equation will be $y = 3x + c$. Using $x = -1$ and $y = -3$ gives:

$-3 = -3 + c \Rightarrow c = 0$

So the equation of BC is $y = 3x$.

The lines intersect when:

$3x = -\frac{1}{3}x + \frac{10}{3}$ and $y = 3 \times 1$

$x = 1$ $\qquad\qquad$ $y = 3$

So the point of intersection is at (1, 3).

- **Use the information about the parallel line to find the gradient of the line** AC.

- You can use the **coordinates of the point A** to find the value of c in the equation.

- As the lines AC and BC are perpendicular, the **product of their gradients must be −1**, so:

gradient of BC = $\dfrac{-1}{\text{gradient of AC}}$

- When the lines intersect, they will have the same y-value. Use this to **form and solve an equation** to find x.

Q2 **(a)** A circle has its centre at (3, 5) and radius $\sqrt{2}$.

  (i) Write down the equation of this circle.
  (ii) Find the coordinates of the point where this circle intersects the line $y = x$.

(b) The equation of a circle is
$x^2 - 6x + y^2 - 4y + 9 = 0$.
Find the centre and radius of this circle.

(a) (i) $(x - 3)^2 + (y - 5)^2 = (\sqrt{2})^2$
$\qquad$ $(x - 3)^2 + (y - 5)^2 = 2$
(ii) $(x - 3)^2 + (x - 5)^2 = 2$
$\qquad$ $2x^2 - 16x + 32 = 0$
$\qquad$ $x^2 - 8x + 16 = 0$
$\qquad$ $(x - 4)^2 = 0$
$\qquad$ $x = 4$ and $y = 4$

(b) $x^2 - 6x + y^2 - 4y + 9 = 0$
$\qquad$ $(x - 3)^2 - 9 + (y - 2)^2 - 4 + 9 = 0$
$\qquad$ $(x - 3)^2 + (y - 2)^2 = 4 = 2^2$
$\qquad$ Centre is (3, 2). Radius 2.

- **Substitute the coordinates of the centre and the radius into the standard equation of the circle** to obtain the required equation.

- To find the intersection point, substitute $y = x$, then simplify and solve the resulting quadratic equation. In this case the equation can be divided by 2 and then factorised. Don't forget to find the y-coordinate.

- The form of the equation must be changed using the ideas of **completing the square**. Note that $x^2 - 6x = (x - 3)^2 - 9$. Once the equation is in the standard format the centre and the radius can be written down.

Q1 The points A and B have coordinates (1, 6) and (5, 14) respectively. Find the equation of the line *p* that passes through the points A and B, and the line *q* that is perpendicular to *p* and passes through the midpoint of AB.

Q2 The lines $y - 3x + 6 = 0$ and $3y + x - 12 = 0$ intersect at the point A.

(a) Find the coordinates of A.

(b) Show that the two lines are perpendicular.

(c) Find the area of the triangle formed by the two lines and the *x*-axis.

Q3 The line *p* is parallel to the line with equation $3y - x + 4 = 0$, and intersects the *x*-axis at (–2, 0). The line *q* is perpendicular to the line *p* and intersects the *x*-axis at (8, 0). Find the equation of each line and the coordinates of their point of intersection.

Q4 The points A, B and C have coordinates (3, 6), (6, 5) and (7, 2) respectively.

(a) Find the equation of the line *p* that passes through the points A and C and the line *q* that is perpendicular to *p* and passes through the point B.

(b) Find the coordinates of the point where the lines *p* and *q* intersect.

Q5 The points A(–2, 4), B(6, –2) and C(5, 5) are the vertices of ΔABC and D is the midpoint of AB.

(a) Find an equation of the line passing through A and B in the form $ax + by + c = 0$, where *a*, *b* and *c* are integers to be found.

(b) Show that CD is perpendicular to AB.

Q6 A circle has centre (4, –2) and radius $\sqrt{8}$.

(a) Write down the equation of the circle.

(b) Find the coordinates of the points where the circle intersects the line with equation $y - x + 6 = 0$.

Q7 The equation of a circle is $x^2 - 14x + y^2 - 10y + 49 = 0$. Find the centre and radius of this circle.

Q8 The points A (10, 5) and B (2, 11) lie on the opposite ends of the diameter of a circle. Find the equation of this circle.

Q9 A circle C has equation $x^2 + y^2 - 12x = 0$.

(a) Find the radius and the coordinates of the centre of the circle C.

(b) Find the coordinates of the points where the line with equation $y = 2x$ intersects the circle.

Answers can be found on pages 128–130.

Key points to remember

- The **binomial expansion** is used to expand expressions of the form $(a \pm b)^n$.

- When n is a positive integer, the expansion contains a finite number of terms as given in the formula below.

$$(a + b)^n = a^n + \binom{n}{1}a^{n-1}b + \binom{n}{2}a^{n-2}b^2 + \ldots + \binom{n}{r}a^{n-r}b^r + \ldots + b^n \qquad n \in \mathbb{N}$$

where $\binom{n}{r} = {}^nC_r = \dfrac{n!}{(n-r)!\, r!}$

or $(1 + x)^n = 1 + nx + \dfrac{n(n-1)x^2}{2!} + \dfrac{n(n-1)(n-2)x^3}{3!} + \ldots$

This can be found in your formula book. Make sure that you know where it is.

- The coefficients $\binom{n}{r}$ can be obtained from Pascal's triangle for small values of n.

```
                    1
               1         1
          1         2         1
     1         3         3         1
1         4         6         4         1
1    5         10        10        5         1
```

The coefficients for expanding a bracket to the power 5 would be 1, 5, 10, 10, 5, 1.

Formulae you must know

- The key formulae for this topic will be included in your formula book. Make sure that you are able to find and use them.

Don't make these mistakes...

Don't forget to consider when an expansion will be valid if n is not a positive integer.

Don't forget to include the '−' signs when there is one in the bracket.

In cases like $(1 + 3x)^n$, don't forget to use $(3x)^2 = 9x^2$ etc. when doing the expansion.

Don't make arithmetic errors when simplifying expansions.

Q1 Expand $(1 + 2x)^4$.

$(1 + 2x)^4 = 1 + 4 \times (2x) + 6 \times (2x)^2 + 4 \times (2x)^3 + (2x)^4$

$\qquad = 1 + 4 \times 2x + 6 \times 4x^2 + 4 \times 8x^3 + 16x^4$

$\qquad = 1 + 8x + 24x^2 + 32x^3 + 16x^4$

- Show clearly how to obtain each term of the expansion. The coefficients 1, 4, 6, 4, 1 can be obtained from **Pascal's triangle**.
- **Simplify** the expression carefully.

Q2 Expand $(2 - 3x)^4$.

$(2 - 3x)^4 = 1 \times 2^4 + 4 \times 2^3 \times (-3x) + 6 \times 2^2 \times (-3x)^2$
$\qquad\qquad + 4 \times 2 \times (-3x)^3 + 1 \times (-3x)^4$

$\qquad = 16 - 96x + 216x^2 - 216x^3 + 81x^4$

- Write down clearly how to obtain each term. Make sure that you **put the** $(-3x)$ **in brackets each time it appears**.
- **Simplify**, being particularly careful with the negative signs.

Q3 Expand $\left(2 + \dfrac{x}{2}\right)^4$.

$\left(2 + \dfrac{x}{2}\right)^4 = 2^4 + 4 \times 2^3 \times \left(\dfrac{x}{2}\right) + 6 \times 2^2 \times \left(\dfrac{x}{2}\right)^2 + 4 \times 2 \times \left(\dfrac{x}{2}\right)^3 + \left(\dfrac{x}{2}\right)^4$

$\qquad = 2^4 + 4 \times 2^3 \times \left(\dfrac{x}{2}\right) + 6 \times 2^2 \times \left(\dfrac{x}{2}\right)^2 + 4 \times 2 \times \left(\dfrac{x}{2}\right)^3 + \left(\dfrac{x}{2}\right)^4$

$\qquad = 16 + 4 \times 8 \times \dfrac{x}{2} + 6 \times 4 \times \dfrac{x^2}{4} + 4 \times 2 \times \dfrac{x^3}{8} + \dfrac{x^4}{16}$

$\qquad = 16 + 16x + 6x^2 + x^3 + \dfrac{x^4}{16}$

- **Use the formula for expanding** $(a + b)^n$ that is quoted above. Note that in this case $a = 2$ and $b = \dfrac{x}{2}$. The whole of the term $\dfrac{x}{2}$ is squared, cubed, etc.
- Care is needed when simplifying the expression.
- It is usual to **write the expansion in increasing powers of** x.

Q4 If $(1 + ax)^n = 1 + 15x + 90x^2 + \ldots$, find a and n.

$(1 + ax)^n = 1 + n(ax) + \dfrac{n(n-1)}{2}(ax)^2 + \ldots$

$\qquad = 1 + nax + \dfrac{n(n-1)}{2}a^2x^2 + \ldots$

The coefficient of x is 15 so

$\qquad an = 15 \qquad$ or $\qquad a = \dfrac{15}{n}$

The coefficient of x^2 is 90 so

$\qquad \dfrac{n(n-1)}{2}a^2 = 90$

Substituting for a gives:

$\qquad \dfrac{n(n-1)}{2} \times \left(\dfrac{15}{n}\right)^2 = 90$

$\qquad\qquad n - 1 = \dfrac{90 \times 2}{15^2}n$

$\qquad\qquad n - 1 = \dfrac{4}{5}n$

$\qquad\qquad \dfrac{1}{5}n = 1$

$\qquad\qquad n = 5$

Then

$\qquad a = \dfrac{15}{n} = \dfrac{15}{5} = 3$

- First **consider the expansion of** $(1 + ax)^n$ using the formula and replacing x by ax.
- **Compare the coefficients of** x in both to form an equation that relates a and n.
- **Compare the coefficients of** x^2 to form a second equation.
- **Substitute** for a in the second equation using $a = \dfrac{15}{n}$ and solve for n.
- Use $n = 5$ to find a.

Q1 Expand $(1 + 2x)^3$.

Q2 Expand $(4 - x)^5$.

Q3 Expand $(5 - 3x)^4$.

Q4 A polynomial p(x) is defined as

$$p(x) = (2 + 3x)^5$$

Use the binomial theorem to find the coefficient of x^4 when p(x) is expanded.

Q5 Expand $\left(x - \dfrac{1}{x}\right)^3$

Q6 Expand $(1 + x)^5$ in ascending powers of x, simplifying the coefficients.

Hence by letting $x = y + y^2$, find the coefficient of y^4 in the expansion of $(1 + y + y^2)^5$ in powers of y.

Q7 If $(1 - ax)^n = 1 - 20x + 160x^2 + \ldots$ find the values of a and n.

Answers can be found on page 130.

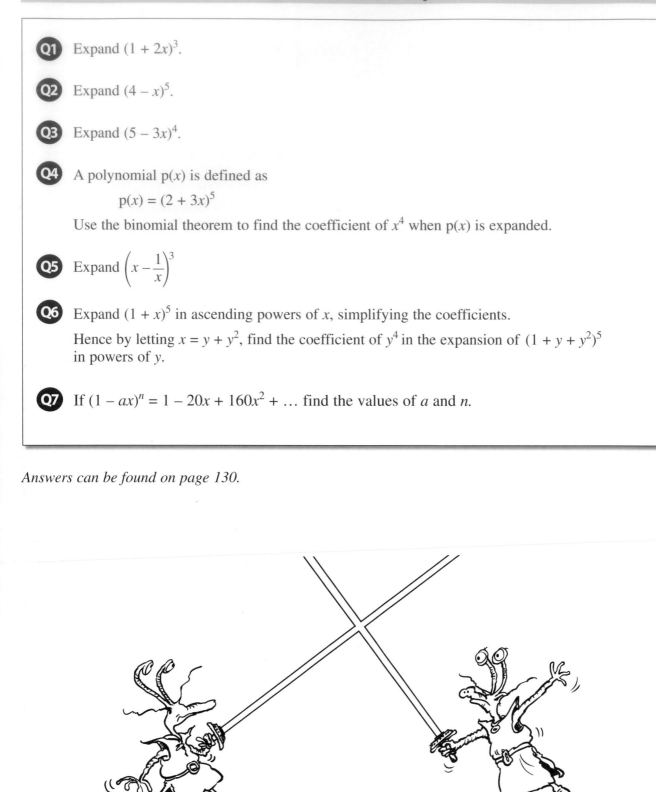

6 Differentiation

Techniques

Simplify brackets and fractions before differentiating.

- Write $(x^2 - 4)x$ as $x^3 - 4x$ before differentiating.

- Write $\dfrac{x^2 + 2}{x}$ as $x + 2x^{-1}$ before differentiating.

Write reciprocals and roots as powers.

- Write $\sqrt{x}$ as $x^{\frac{1}{2}}$ and $\dfrac{1}{x^3}$ as x^{-3} before differentiating.

Stationary points

At a stationary point $\dfrac{dy}{dx} = 0$.

A stationary point can be a **local maximum**, a **local minimum** or a **point of inflexion**. Test by considering the gradient either side of the stationary point.

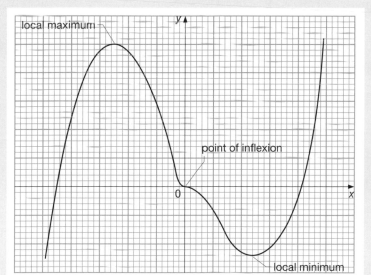

If $\dfrac{d^2y}{dx^2} < 0$ at a stationary point, it will be a local maximum.

If $\dfrac{d^2y}{dx^2} > 0$ at a stationary point, it will be a local minimum.

If $\dfrac{d^2y}{dx^2} = 0$ at a stationary point, determine the nature by calculating the gradient either side of the stationary point.

Gradients of curves

The **gradient of a curve** is given by its **derivative**.

Tangents and normals

The gradient of a tangent will be given by $\dfrac{dy}{dx}$.

The gradient of a normal will be given by $-1 \Big/ \dfrac{dy}{dx}$.

Increasing and decreasing functions

A function $f(x)$ is **increasing** on the interval (a, b), if $f'(x) > 0$, for all x in (a, b).

A function $f(x)$ is **decreasing** on the interval (a, b), if $f'(x) < 0$, for all x in (a, b).

Formulae you must know

- Derivative of x^n

  $$\frac{d}{dx}(x^n) = nx^{n-1}$$

- Function notation

  If $y = f(x)$ then $\dfrac{dy}{dx} = f'(x)$

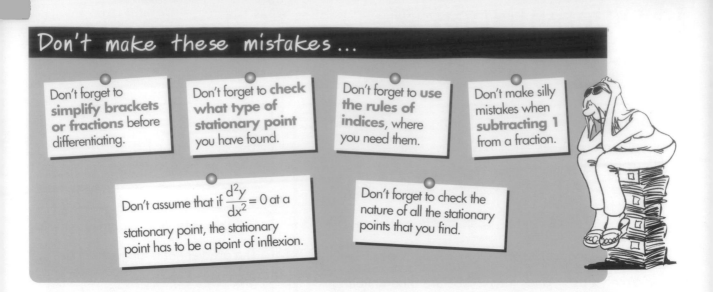

Don't make these mistakes...

Don't forget to **simplify brackets or fractions** before differentiating.

Don't forget to **check what type of stationary point** you have found.

Don't forget to **use the rules of indices**, where you need them.

Don't make silly mistakes when **subtracting 1** from a fraction.

Don't assume that if $\dfrac{d^2y}{dx^2} = 0$ at a stationary point, the stationary point has to be a point of inflexion.

Don't forget to check the nature of all the stationary points that you find.

Exam Questions and Student's Answers

Q1 If $y = x^2\left(3x - 1 + \dfrac{1}{x}\right)$ find $\dfrac{dy}{dx}$.

$$y = x^2\left(3x - 1 + \dfrac{1}{x}\right) = 3x^3 - x^2 + x$$

So $\dfrac{dy}{dx} = 9x^2 - 2x + 1$

Q2 If $f(x) = \dfrac{\sqrt{x} + 2}{x^2}$ find $f'(x)$.

$$f(x) = \dfrac{\sqrt{x} + 2}{x^2}$$

$$= \dfrac{\sqrt{x}}{x^2} + \dfrac{2}{x^2}$$

$$= \dfrac{x^{\frac{1}{2}}}{x^2} + \dfrac{2}{x^2}$$

$$= x^{-\frac{3}{2}} + 2x^{-2}$$

So $f'(x) = -\dfrac{3}{2}x^{-\frac{3}{2} - 1} - 4x^{-2 - 1}$

$$= -\dfrac{3}{2}x^{-\frac{5}{2}} - 4x^{-3}$$

$$= -\dfrac{3}{2\sqrt{x^5}} - \dfrac{4}{x^3}$$

How to score full marks

- **Expand the brackets** before differentiating.

- Remember to **multiply each term inside the bracket** by x^2.

- You must **simplify** each term before differentiating.

- Differentiate each term using the rule for x^n.

- You don't know how to **differentiate a fraction** like this, so you must **simplify it first**.

- First **split the fraction into two parts**.

- Then write them in the form x^n.

- Now you can differentiate using **the rule for differentiating x^n**.

- You can also write the result in this form.

Q3 Find the equation of the normal to the curve $y = x + \dfrac{4}{x^2}$, at the point with coordinates (1, 5).

$y = x + 4x^{-2}$

$\dfrac{dy}{dx} = 1 - 8x^{-3} = 1 - \dfrac{8}{x^3}$

If $x = 1$

$\dfrac{dy}{dx} = 1 - \dfrac{8}{1^3} = -7$

Gradient of normal $= \dfrac{-1}{-7} = \dfrac{1}{7}$

Equation of normal is:

$y = \dfrac{1}{7}x + c$

$x = 1, y = 5$

$5 = \dfrac{1}{7} + c$

$c = \dfrac{34}{7}$

$y = \dfrac{1}{7}x + \dfrac{34}{7}$

- To find the gradient of the normal you need to **first find** $\dfrac{dy}{dx}$.

- Find the value of $\dfrac{dy}{dx}$ when $x = 1$.

- The gradient of the normal is given by $-1 \Big/ \dfrac{dy}{dx}$.

- The equation of the normal will be of the form $y = mx + c$.

- The coordinates of the point on the curve can be used to find the value of c.

I THOUGHT I'D REACHED THE TOP-BUT IT'S ONLY A LOCAL MAXIMUM!

Q4 Determine the coordinates of the stationary points of the curve with equation $y = \dfrac{2}{x} + 8x$.

Also determine whether each of these points is a local maximum or minimum.

$y = 2x^{-1} + 8x$

$\dfrac{dy}{dx} = -2x^{-2} + 8$

For stationary points $\dfrac{dy}{dx} = 0$.

$0 = -\dfrac{2}{x^2} + 8$

$x^2 = \dfrac{1}{4}$

$x = \dfrac{1}{2}$ or $x = -\dfrac{1}{2}$

x	$\dfrac{1}{4}$	$\dfrac{1}{2}$	1
$\dfrac{dy}{dx}$	-24	0	6

There is a local minimum at $(\frac{1}{2}, 8)$.

x	-1	$-\dfrac{1}{2}$	$-\dfrac{1}{4}$
$\dfrac{dy}{dx}$	6	0	-24

There is a local maximum at $(-\frac{1}{2}, -8)$.

Alternative approach

$\dfrac{d^2y}{dx^2} = 4x^{-3} = \dfrac{4}{x^3}$

When $x = \frac{1}{2}$, $\dfrac{d^2y}{dx^2} = 32$, which is greater than zero so a local minimum at $(\frac{1}{2}, 8)$.

When $x = -\frac{1}{2}$, $\dfrac{d^2y}{dx^2} = -32$, which is less than zero so a local maximum at $(-\frac{1}{2}, -8)$.

How to score full marks

- First write $\dfrac{2}{x}$ as $2x^{-1}$, so that it can be **differentiated**.

- Remember that at **stationary points the derivative is zero**.
- Solve the equation that you form, **finding both values of x**.

- **Substitute values for x either side of $x = \frac{1}{2}$ to find the gradients.** Here the gradient changes from being negative to positive at the stationary point so it is a local minimum. **Don't forget to calculate the y-coordinate as well.**

- For the negative value of x, the gradient changes from being positive to negative, indicating that there is a **local maximum**.

- Rather than using tables it is possible to **consider the second derivative.**

- Once this has been obtained the two x-values can be substituted to **determine the nature of the stationary points.**

Q1 Given that $y = x^4 - 3x^2 - x - 2$, find $\dfrac{dy}{dx}$.

Q2 If $y = x^3 + 6x - 8$ find $\dfrac{dy}{dx}$.

Q3 If $f(x) = \sqrt{x}\left(x + \dfrac{2}{x}\right)$, find $f'(x)$.

Q4 If $f(x) = x^2 + \dfrac{16}{x}$, find the range of positive values of x for which $f(x)$ is increasing.

Q5 A curve has equation $y = (x^2 + 6)(x - 3)$.

(a) Find $\dfrac{dy}{dx}$.

(b) Determine the values of x for which the gradient of the curve is equal to 15.

Q6 The equation of a curve is $y = 9x^2 - 4x^3$. Find the coordinates of the two stationary points on the curve, and determine the nature of these stationary points.

Determine the set of values of x for which $9x^2 - 4x^3$ is a decreasing function of x.

Q7 Find the equation of the tangent and the normal to the curve $y = 4x - \sqrt{x}$, at the point with coordinates $(4, 14)$.

Q8 A builder wishes to construct a rectangular storage area. He has 20m of fencing available and must have a gate of width 5 m in the middle of one side of the rectangle. The opposite side of the rectangle is an existing wall.

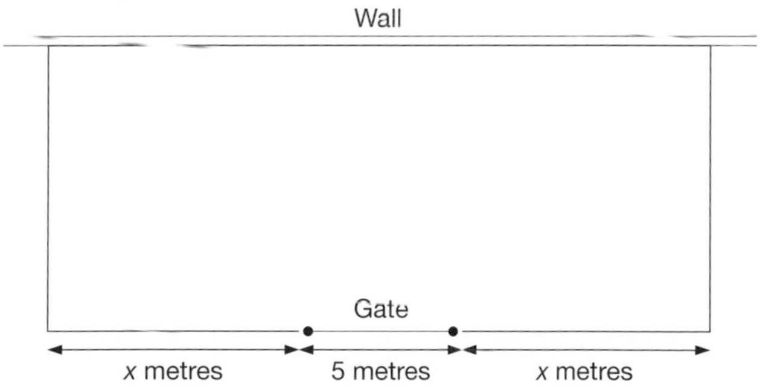

He constructs the storage area so that there are x metres of fencing each side of the gate, as shown in the diagram. He uses all 20 m of the fencing available. The area enclosed is $A\,\text{m}^2$.

(a) Show that $A = (2x + 5)(10 - x)$.

(b) Find the maximum value of A, giving your answer correct to three significant figures.

Q9 An architect is drawing up plans for a mini-theatre. The diagram shows the plan of the base which consists of a rectangle of length $2y$ metres and width $2x$ metres and a semicircle of radius x metres which is placed with one side of the rectangle as diameter.

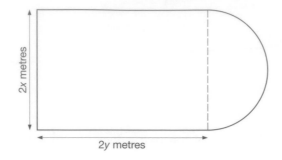

Find, in terms of x and y, expressions for:

(a) the perimeter of the base

(b) the area of the base.

The architect decides that the base should have a perimeter of 100 metres.

(c) Show that the area, A square metres, of the base is given by:
$$A = 100x - 2x^2 - \tfrac{1}{2}\pi x^2$$

(d) Given that x can vary, find the value of x for which $\dfrac{dA}{dx} = 0$ and determine the corresponding value of y, giving your answer to two significant figures.

(e) Find the maximum value of A and explain why this is a maximum.

Q10 The curve C is defined by the equation $y = 2x^2\sqrt{x} + \dfrac{1}{x^4}$ for $x > 0$.

(a) Write $x^2\sqrt{x}$ in the form x^k, where k is a fraction.

(b) Find $\dfrac{dy}{dx}$.

(c) Find an equation of the tangent to the curve C at the point on the curve where $x = 1$.

(d) (i) Find $\dfrac{d^2y}{dx^2}$.

(ii) Hence deduce that the curve C has no maximum points.

Q11 **(a)** Find the coordinates of the stationary points on the curve $y = 2x^3 - 3x^2 - 12x - 7$.

(b) Determine whether each stationary point is a maximum point or a minimum point.

(c) By expanding the right hand side, show that $2x^3 - 3x^2 - 12x - 7 = (x + 1)^2 (2x - 7)$.

(d) Sketch the curve $y = 2x^3 - 3x^2 - 12x - 7$, marking the coordinates of the stationary points and the points where the curve meets the axes.

Answers can be found on pages 131–132.

Key points to remember

● **Areas under curves**

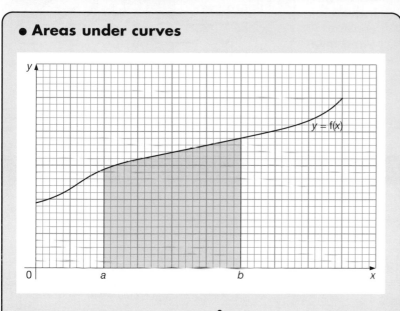

Area of shaded region $= \int_a^b f(x)dx.$

Where an integral is used to find an area under the x-axis, the result will be **negative**.

For example consider the curve $y = x^2 - 4$, as shown in the diagram.

$\int_{-2}^{2}(x^2 - 4)dx = -\frac{32}{3}$

$\int_{2}^{4}(x^2 - 4)dx = \frac{32}{3}$

$\int_{-2}^{4}(x^2 - 4)dx = 0$

But the area of the shaded region is

$\frac{32}{3} + \frac{32}{3} = \frac{64}{3}$

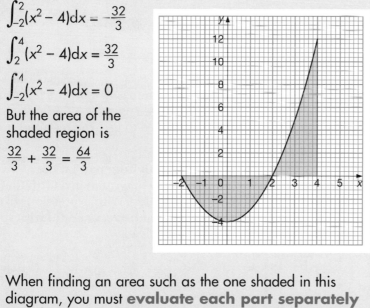

When finding an area such as the one shaded in this diagram, you must **evaluate each part separately** and change the sign on any negative parts before calculating the total area.

● **Definite integrals**

These are integrals such as $\int_0^2 x^2 dx$, which are evaluated to give a **specific value**.

● **Indefinite integrals**

These are integrals such as $\int x^2 dx$, which will be in terms of x and must include a **constant of integration**, c.

THE AREA UNDER THIS CURVE IS POSITIVELY DRY!

Formulae you must know

● Standard integral

$\int x^n dx = \frac{x^{n+1}}{n+1} + c$ if $n \neq -1$

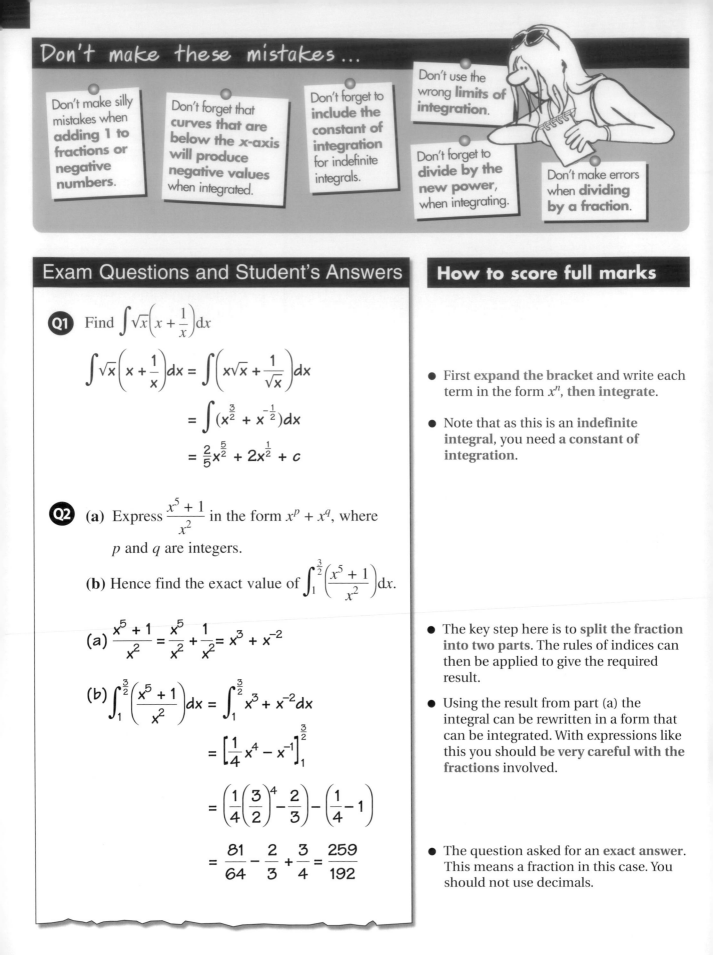

Don't make silly mistakes when **adding 1 to fractions or negative numbers**.

Don't forget that **curves that are below the x-axis will produce negative values** when integrated.

Don't forget to **include the constant of integration** for indefinite integrals.

Don't use the **wrong limits of integration**.

Don't forget to **divide by the new power**, when integrating.

Don't make errors when **dividing by a fraction**.

Exam Questions and Student's Answers

How to score full marks

Q1 Find $\int \sqrt{x}\left(x + \dfrac{1}{x}\right)dx$

$$\int \sqrt{x}\left(x + \frac{1}{x}\right)dx = \int \left(x\sqrt{x} + \frac{1}{\sqrt{x}}\right)dx$$

$$= \int \left(x^{\frac{3}{2}} + x^{-\frac{1}{2}}\right)dx$$

$$= \frac{2}{5}x^{\frac{5}{2}} + 2x^{\frac{1}{2}} + c$$

- First **expand the bracket** and write each term in the form x^n, **then integrate**.

- Note that as this is an **indefinite integral**, you need **a constant of integration**.

Q2 **(a)** Express $\dfrac{x^5 + 1}{x^2}$ in the form $x^p + x^q$, where p and q are integers.

(b) Hence find the exact value of $\int_1^{\frac{3}{2}}\left(\dfrac{x^5 + 1}{x^2}\right)dx$.

(a) $\dfrac{x^5 + 1}{x^2} = \dfrac{x^5}{x^2} + \dfrac{1}{x^2} = x^3 + x^{-2}$

(b) $\int_1^{\frac{3}{2}}\left(\dfrac{x^5 + 1}{x^2}\right)dx = \int_1^{\frac{3}{2}} x^3 + x^{-2}dx$

$$= \left[\frac{1}{4}x^4 - x^{-1}\right]_1^{\frac{3}{2}}$$

$$= \left(\frac{1}{4}\left(\frac{3}{2}\right)^4 - \frac{2}{3}\right) - \left(\frac{1}{4} - 1\right)$$

$$= \frac{81}{64} - \frac{2}{3} + \frac{3}{4} = \frac{259}{192}$$

- The key step here is to **split the fraction into two parts**. The rules of indices can then be applied to give the required result.

- Using the result from part (a) the integral can be rewritten in a form that can be integrated. With expressions like this you should **be very careful with the fractions** involved.

- The question asked for an **exact answer**. This means a fraction in this case. You should not use decimals.

Q3 **(a)** State the solutions of the equation
$(x - 1)^2 (x - 5) = 0$.

(b) Find the area of the finite region enclosed
by the x-axis and the curve with equation
$y = (x - 1)^2 (x - 5)$.

(a) $(x - 1)^2 (x - 5) = 0$

$x - 1 = 0$ or $x - 5 = 0$

$x = 1$ or $x = 5$

(b) $(x - 1)^2(x - 5) = (x^2 - 2x + 1)(x - 5)$

$= x^3 - 2x^2 + x - 5x^2 + 10x - 5$

$= x^3 - 7x^2 + 11x - 5$

$\int_1^5 (x^3 - 7x^2 + 11x - 5)dx$

$= \left[\dfrac{1}{4}x^4 - \dfrac{7}{3}x^3 + \dfrac{11}{2}x^2 - 5x \right]_1^5$

$= \left(\dfrac{625}{4} - \dfrac{875}{3} + \dfrac{275}{2} - 25 \right) - \left(\dfrac{1}{4} - \dfrac{7}{3} + \dfrac{11}{2} - 5 \right)$

$= \left(-\dfrac{275}{12} \right) - \left(-\dfrac{19}{12} \right) = -\dfrac{64}{3}$

Area $= \dfrac{64}{3}$

Q4 The points A and B lie on the curve $y = 5 + 2x - x^2$
and have coordinates $(0, 5)$ and $(3, 2)$
respectively. Find the area of the region enclosed
by the curve and the line AB.

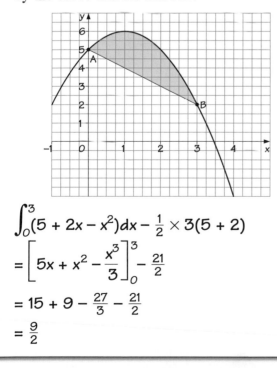

$\int_0^3 (5 + 2x - x^2)dx - \dfrac{1}{2} \times 3(5 + 2)$

$= \left[5x + x^2 - \dfrac{x^3}{3} \right]_0^3 - \dfrac{21}{2}$

$= 15 + 9 - \dfrac{27}{3} - \dfrac{21}{2}$

$= \dfrac{9}{2}$

- To find the solutions of the equation, **set
each of the factors equal to zero**, which
gives the two solutions.

- First **expand the brackets**, so that the
equation is in a form that can be
integrated.

- **The limits of integration** will be 1 and 5,
because these are the points at which
the curve intersects the x-axis. The
actual integral is **negative**. This is
because the **enclosed area is below the
x-axis**, as shown in the graph below.

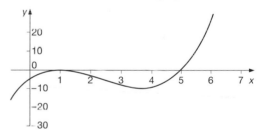

- It **is important** that you draw a diagram
to **identify the area** that is required.

- The shaded area is given by
$\int_0^3 (5 + 2x - x^2)dx$ less the area of the
trapezium with corners at $(0, 0)$, $(3, 0)$,
$(3, 2)$ and $(0, 5)$. So you can evaluate the
area as shown.

- Note the use of the formula
$A = \frac{1}{2}h(a + b)$ **for the area of the
trapezium**.

Q1 Find $\int_0^3 (2x^3 - x^2)\,dx$.

Q2 Find $\int \dfrac{(x^2 - x)}{\sqrt{x}}\,dx$.

Q3 Find the area of the region bounded by the curve $y = x^2$, the line $y = 12 - x$ and the x-axis.

Q4 Find the area of the region bounded by the curve $y = x^2 - 5x + 9$ and the line $y = 3$.

Q5 The points A and B lie on the curve $y = 16 - x^4$ and have coordinates $(-2, 0)$ and $(1, 15)$ respectively. Find the area of the finite region enclosed by the curve and the line AB.

Q6 The points A and B lie on the curve $y = x^2(5 - x)$ and have coordinates $(1, 4)$ and $(4, 16)$ respectively. Find the area of the region enclosed by the curve and the chord AB.

Q7 The graph on the right shows the curve with equation $y = x^3 - 5x^2 + 6x$. Find the total area of the regions that have been shaded on the diagram.

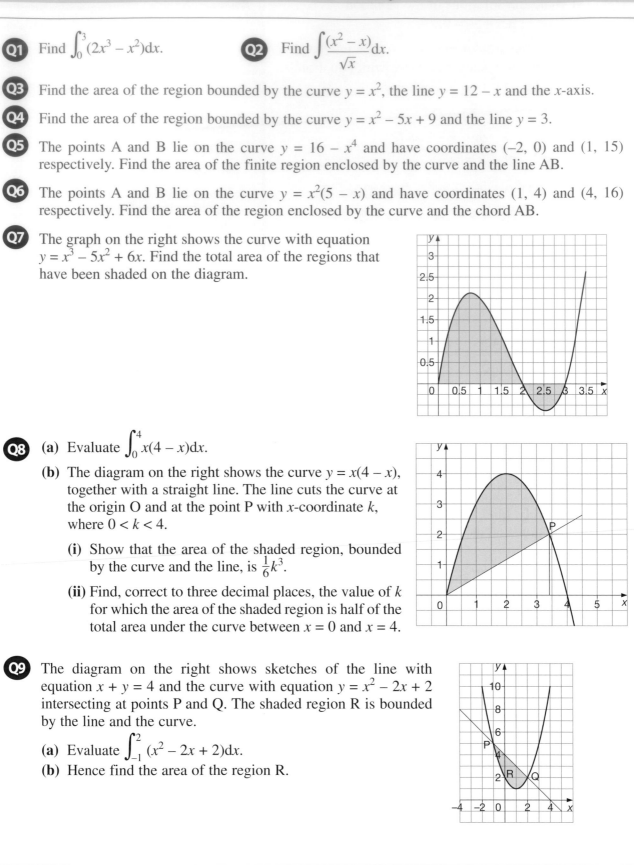

Q8 (a) Evaluate $\int_0^4 x(4 - x)\,dx$.

(b) The diagram on the right shows the curve $y = x(4 - x)$, together with a straight line. The line cuts the curve at the origin O and at the point P with x-coordinate k, where $0 < k < 4$.

   (i) Show that the area of the shaded region, bounded by the curve and the line, is $\frac{1}{6}k^3$.

   (ii) Find, correct to three decimal places, the value of k for which the area of the shaded region is half of the total area under the curve between $x = 0$ and $x = 4$.

Q9 The diagram on the right shows sketches of the line with equation $x + y = 4$ and the curve with equation $y = x^2 - 2x + 2$ intersecting at points P and Q. The shaded region R is bounded by the line and the curve.

(a) Evaluate $\int_{-1}^2 (x^2 - 2x + 2)\,dx$.

(b) Hence find the area of the region R.

Answers can be found on pages 132–134.

Key points to remember

- The **trapezium rule** is a method for **estimating the area under a curve** by finding the sum of the areas of a number of trapeziums.

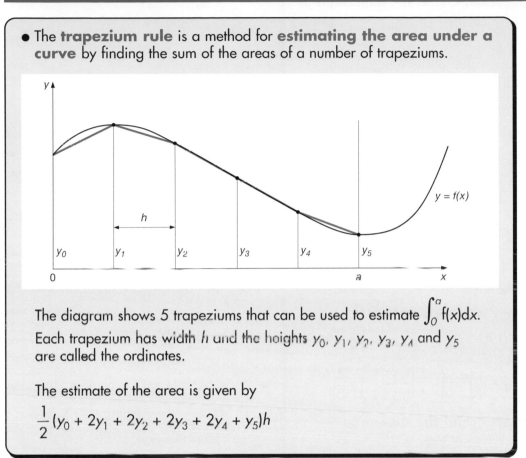

The diagram shows 5 trapeziums that can be used to estimate $\int_0^a f(x)dx$.

Each trapezium has width h and the heights y_0, y_1, y_2, y_3, y_4 and y_5 are called the ordinates.

The estimate of the area is given by

$$\frac{1}{2}(y_0 + 2y_1 + 2y_2 + 2y_3 + 2y_4 + y_5)h$$

Formulae you must know

- $\int_0^a f(x)dx \approx \frac{1}{2}(y_0 + 2y_1 + 2y_2 + \ldots + 2y_{n-1} + y_n)h$

Don't make these mistakes ...

Don't double the first or final ordinate.

Don't make silly arithmetic errors when evaluating integrals.

Don't forget to multiply by h.

Exam Questions and Student's Answers

Q1 (a) Use the trapezium rule with 4 trapeziums to estimate $\int_1^3 \frac{1}{x^2}\,dx$.

(b) Find the exact value of the integral and compare with your answer to part (a).

(c) State how your answer to part (a) could be improved.

(a) $h = \frac{3-1}{4} = 0.5$

x	1	1.5	2	2.5	3
$\frac{1}{x^2}$	1	0.4444	0.25	0.16	0.1111

$$\int_1^3 \frac{1}{x^2}dx \approx \frac{1}{2}(1 + 2\times 0.4444 + 2\times 0.25 + 2\times 0.16 + 0.1111)\times 0.5$$
$$= 0.70$$

(b) $\int_1^3 \frac{1}{x^2}dx = \left[-\frac{1}{x}\right]_1^3 = \left(-\frac{1}{3}\right) - \left(-\frac{1}{1}\right) = \frac{2}{3}$

This is smaller than the answer to part (a).

(c) The estimate could be improved by using a smaller h, which means that you fit thinner trapeziums into the same space.

Q2 Use 5 trapeziums to estimate the value of $\int_0^1 0.5^x dx$.

$h = \frac{1-0}{5} = 0.2$

x	0	0.2	0.4	0.6	0.8	1
0.5^x	1	0.8706	0.7579	0.6598	0.5743	0.5

$$\int_0^1 0.5^x dx \approx \frac{1}{2}(1 + 2\times 0.8706 + 2\times 0.7579 + 2\times 0.6598 + 2\times 0.5743 + 0.5)\times 0.2$$
$$= 0.72$$

How to score full marks

- Calculate the **step size** based on the limits of the integral and the number of trapeziums.
- Calculate the values of the **ordinates**. It is a good idea to list them in a tabular format.
- **The formula can then be used** to calculate the estimate of the integral.
- The exact integral can be found by using $\int x^n dx = \frac{x^{n+1}}{n+1} + c$ where $n = -2$.
- **Estimates can be improved** by using **more trapeziums**.
- Calculate the value of h.
- Calculate the values of the **ordinates** for each value of x.
- Then apply the **trapezium rule formula**.
- Note that the answer is given to a small number of significant figures.

Questions to try

Q1 Complete the table below and use it to estimate the value of $\int_0^1 2^x \, dx$.

x	0	0.25	0.5	0.75	1
2^x					

Q2 Use the trapezium rule with the table below to estimate $\int_1^2 \frac{2}{x} \, dx$.

x	1	1.2	1.4	1.6	1.8	2.0
$\frac{2}{x}$						

Q3 Use the trapezium rule with 5 trapeziums to estimate the value of $\int_1^2 x + \frac{1}{x} \, dx$.

Q4 The integral $\int_1^3 \frac{1}{\sqrt{x}} \, dx$ is to be evaluated. Give your answers correct to two decimal places.

(a) Use the trapezium rule with 2 trapeziums to estimate the value of this integral.

(b) Use the trapezium rule with 4 intervals to estimate the value of the integral.

(c) Evaluate the integral exactly.

(d) Explain why the second estimate is better.

Answers can be found on page 134.

9 Factor and remainder theorems

Key points to remember

● **Factor theorem**

If $f(a) = 0$, then $(x - a)$ is a factor of the polynomial $f(x)$.

● **Remainder theorem**

If $f(a) = k$, then k is the remainder when $f(x)$ is divided by $(x - a)$.

● **Algebraic long division**

$$
\begin{array}{r}
x^2 - x - 6 \\
x - 1\overline{)\, x^3 - 2x^2 - 5x + 6} \\
\underline{x^3 - x^2} \\
-x^2 - 5x \\
\underline{-x^2 + x} \\
-6x + 6 \\
\underline{-6x + 6} \\
0
\end{array}
$$

Hence $x^3 - 2x^2 - 5x + 6 = (x - 1)(x^2 - x - 6)$

Don't make these mistakes...

If $f(a) = 0$ then don't assume that $(x + a)$ is a factor of $f(x)$ instead of $(x - a)$.

Don't make silly mistakes with your negative signs when doing algebraic long division.

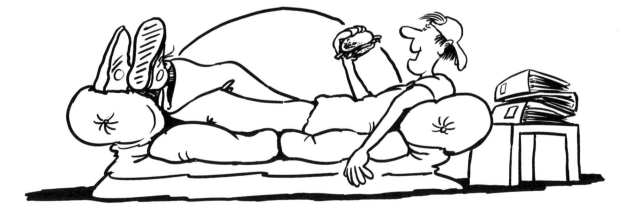

Exam Questions and Student's Answers

Q1 The polynomial f(x) is defined as

$$f(x) = x^3 - 6x^2 + 3x + c$$

(a) Given that $(x - 5)$ is a factor of f(x), find the value of c.

(b) Find the remainder when f(x) is divided by $(x - 4)$.

(a) $f(5) = 0$

$125 - 150 + 15 + c = 0$

$-10 + c = 0$

$c = 10$

(b) $f(4) = 64 - 96 + 12 + 10 = -10$

Remainder $= -10$

Q2 A polynomial is $f(x) = x^3 + 4x^2 - 11x - 30$.

(a) Show that $f(3) = 0$.

(b) Write f(x) as the product of three linear factors.

(c) Hence state the solutions of the equation $f(x) = 0$.

(a) $f(3) = 27 + 36 - 33 - 30 = 0$

(b)

$$\begin{array}{r} x^2 + 7x + 10 \\ x - 3 \,\overline{)\, x^3 + 4x^2 - 11x - 30} \\ x^3 - 3x^2 \qquad\qquad \\ \hline 7x^2 - 11x \qquad \\ 7x^2 - 21x \qquad \\ \hline 10x - 30 \\ 10x - 30 \\ \hline 0 \end{array}$$

$x^2 + 7x + 10 = (x + 2)(x + 5)$

$f(x) = (x + 2)(x + 5)(x - 3)$

(c) $f(x) = 0$

$x = -2$ or $x = -5$ or $x = 3$

How to score full marks

- **As $(x - 5)$ is a factor, then $f(5) = 0$.** This leads to an equation that enables the value of c to be found.
- To find the **remainder** when f(x) is divided by $(x - 4)$, **evaluate f(4)**.
- Simply **substitute $x = 3$** and check that you get zero.
- **Because $f(3) = 0$, then $(x - 3)$ must be a factor.** You can then divide f(x) by $(x - 3)$.
- The result will be a quadratic, which you can then factorise.
- Don't forget to write the result as the **product of three factors** to give the required answer.
- The solutions of $f(x) = 0$ are given by setting each of the factors equal to zero.

Q3 A polynomial is defined as

$$f(x) = x^3 - 4x^2 + ax + b$$

where a and b are positive constants. If $(x + 1)$ is a factor of the polynomial and the remainder is 4 when $f(x)$ is divided by $(x - 1)$, find the values of a and b.

$f(-1) = 0$

$-1 - 4 - a + b = 0$

$a - b = -5$ (1)

$f(1) = 4$

$1 - 4 + a + b = 4$

$a + b = 7$ (2)

$2a = 2$

$a = 1$

$2b = 12$

$b = 6$

- Since $(x + 1)$ is a factor $f(-1) = 0$. This gives one equation.

- Since the remainder is 4 when $f(x)$ is divided by $(x - 1)$, then $f(1) = 4$. This gives a second equation.

- These **simultaneous equations** can now be solved to find a and b.

- **Adding** equations (1) and (2) gives a.

- **Subtracting** equation (1) from equation (2) gives b.

Q1 Divide $x^3 - 5x^2 - 2x + 24$ by $(x - 4)$.

Q2 (a) Divide $x^3 + 3x^2 - 6x - 8$ by $(x + 1)$.

(b) Hence solve $x^3 + 3x^2 - 6x - 8 = 0$.

Q3 (a) Show that $(x + 3)$ is a factor of $x^3 - 3x^2 - 25x - 21$.

(b) Write $x^3 - 3x^2 - 25x - 21$ as the product of three linear factors.

(c) Hence write down the solutions of the equation $x^3 - 3x^2 - 25x - 21 = 0$.

Q4 (a) Given that $(x - 1)$ is a factor of f(x), where f$(x) = x^3 - 4x^2 - kx + 10$, show that $k = 7$.

(b) Divide f(x) by $(x - 1)$ to find a quadratic factor of f(x).

(c) Write f(x) as the product of three linear factors.

(d) Find the value of f(2) and hence write down the remainder when f(x) is divided by $(x - 2)$.

Q5 The polynomial f(x) is defined as

$$f(x) = x^3 - ax^2 + 23x + 36$$

(a) Given that $(x - 4)$ is a factor of f(x), find the value of a.

(b) Find the remainder when f(x) is divided by $(x - 2)$.

(c) Show that $(x - 9)$ is a factor of f(x).

Q6 A polynomial is f$(x) = x^3 - 9x^2 + 14x + 24$.

(a) Show that $(x + 1)$ is a factor of f(x).

(b) Write f(x) as the product of three linear factors.

(c) Hence state the solutions of the equation f$(x) = 0$.

Q7 A polynomial is defined as f$(x) = x^3 - px^2 - qx + 14$, where p and q are positive constants. If $(x - 1)$ is a factor of the polynomial and the remainder is -20 when f(x) is divided by $(x - 2)$, find the values of p and q.

Q8 The cubic polynomial $x^3 + ax^2 + bx - 6$ is denoted by f(x).

(a) The remainder when f(x) is divided by $(x - 2)$ is equal to the remainder when f(x) is divided by $(x + 2)$. Show that $b = -4$.

(b) Given also that $(x - 1)$ is a factor of f(x), find the value of a.

(c) With these values of a and b, express f(x) as the product of a linear factor and a quadratic factor.

(d) Hence determine the number of roots of the equation f$(x) = 0$, explaining your reasoning.

Answers can be found on pages 135–136.

Key points to remember

• **Definition of logarithms**

$\log_a a^n = n$

• **Rules of logarithms**

$\log a + \log b = \log ab$

$\log a - \log b = \log \dfrac{a}{b}$

$n\log a = \log a^n$

• **Graphs of exponentials**

Exponential growth

For example $y = 3^x$

Exponential decay

For example $y = \left(\dfrac{1}{2}\right)^x$

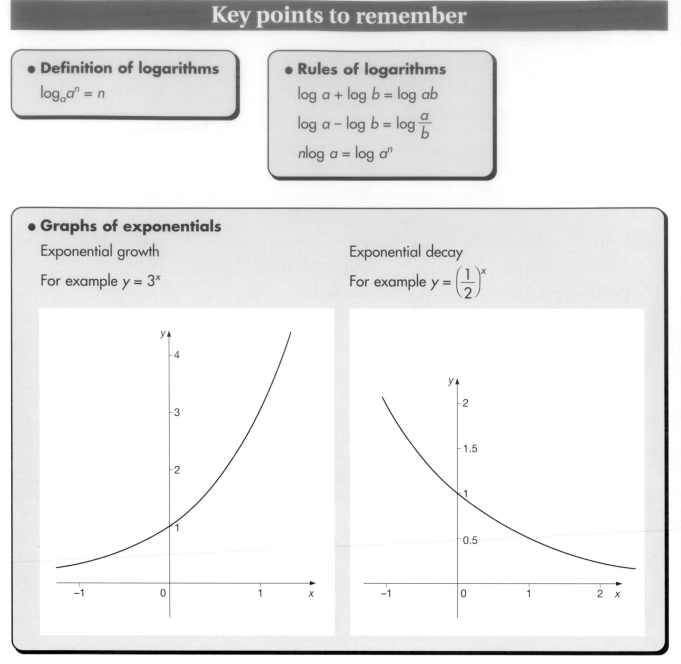

• **Using logarithms to solve equations**

To solve equations like $a = b^x$, take logs of both sides.

$\log a = x \log b$

$x = \dfrac{\log a}{\log b}$

Formulae you must know

• $\log a + \log b = \log ab$

• $\log a - \log b = \log \dfrac{a}{b}$

• $n\log a = \log a^n$

Don't make these mistakes...

$\log(u + v)$ **is not equal to** $\log u + \log v$.

Don't forget the base of the logarithms you are working with.

Don't forget that $\log 1 = 0$ for all bases.

Exam Questions and Student's Answers

Q1 **(a)** Express each of the following in terms of $\log_2 x$:

(i) $\log_2(x^2)$

(ii) $\log_2(8x^2)$

(b) Given that $y^2 = 27$, find the value of $\log_3 y$.

(a) (i) $\log_2(x^2) = 2\log_2 x$

(ii) $\log_2(8x^2) = \log_2 8 + \log_2(x^2)$

$= \log_2(2^3) + 2\log_2 x$

$= 3 + 2\log_2 x$

(b) $\log_3 y = \log_3\sqrt{y^2}$

$= \frac{1}{2}\log_3 y^2$

$= \frac{1}{2}\log_3 27$

$= \frac{1}{2}\log_3 3^3$

$= \frac{3}{2}$

How to score full marks

- For the first part you simply need to use the rule $\log(a^n) = n\log a$.

- First apply the rule $\log(ab) = \log a + \log b$ to split the log into two terms. You can then use the last result and the fact that $8 = 2^3$ to get the required result.

- In this part you need to introduce a y^2 term.

- Remember that $\sqrt{a} = a^{\frac{1}{2}}$.

- Note that $27 = 3^3$.

43

Q2 (a) Given that $\log_a x = \log_a 5 + 2\log_a 3$, where a is a positive constant, show that $x = 45$.

(b) (i) Write down the value of $\log_2 2$.

(ii) Given that $\log_2 y = \log_4 2$, find the value of y.

(a) $\log_a x = \log_a 5 + 2\log_a 3$

$\qquad = \log_a 5 + \log_a 3^2$

$\qquad = \log_a 5 + \log_a 9$

$\qquad = \log_a(5 \times 9) = \log_a 45$

$\quad x = 45$

- This problem requires the use of first the rule $n\log a = \log(a^n)$ and then the rule $\log a + \log b = \log(a \times b)$.

(b) (i) $\log_2 2 = \log_2 2^1 = 1$

- Note that $2 = 2^1$.

(ii) $\log_4 2 = \log_4 \sqrt{4} = \log_4 4^{\frac{1}{2}} = \dfrac{1}{2}$

$\quad \log_2 y = \dfrac{1}{2}$

$\quad y = 2^{\frac{1}{2}} = \sqrt{2}$

- First obtain a value for the right-hand side of the expression, by noting that $2 = \sqrt{4}$.

- Remember that if $p = \log_a q$, then $q = a^p$.

Q3 The value, P, of an investment at time t is given by $P = 3000 \times 1.24^t$.

(a) Find P when $t = 20$.

(b) Find t when $P = 3600$.

(a) $P = 3000 \times 1.24^{20} = 221592$

- Simply substitute $t = 20$ and use your calculator.

(b) $3600 = 3000 \times 1.24^t$

$\quad 1.2 = 1.24^t$

$\quad \log 1.2 = t\log 1.24$

$\quad t = \dfrac{\log 1.2}{\log 1.24} = 0.848$ (to 3 s.f.)

- First substitute $P = 3600$ into the formula. Rearrange until you have the term 1.24^t on one side. Then **take logs of both sides** and solve for t.

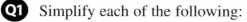

Q1 Simplify each of the following:

(a) $\log_{10}1000000$

(b) $\log_3 81$

(c) $\log_5 0.2$

(d) $\log_4 2$

Q2 If $\log_n a = 4$, find the value of:

(a) $\log_n(a^3)$

(b) $\log_n(\sqrt{a})$

(c) $\log_n\left(\dfrac{1}{a^2}\right)$

Q3 Solve the equation $2 = 1.4^x$.

Q4 Determine the value of n in each of the following statements.

(a) $\log_n 32 = 5$

(b) $\log_5 n = -2$

(c) $\log_3 729 = n$

Q5 Are the following statements true or false?

(a) $\log_4 16 = \log_7 49$

(b) $\dfrac{1}{3}\log_2 8 = \dfrac{1}{3}\log_4 256$

(c) $\log_3 9 = \log_{10}1000 \times \log_4(\sqrt[3]{16})$

Q6 Solve the following equations.

(a) $\log_2 x = \log_5 125 + \log_5 0.008$

(b) $\log_n x + \log_n 15 = \log_n 900$

(c) $2\log_n x - \log_n x = \log_n(2 - x)$

(d) $\log_n x^3 + \log_n x^2 = 10\log_n 5$

Q7 The temperature, T, of an object at time t is given by $T = 80 \times 0.95^t + 20$.

(a) Find T when $t = 0$.

(b) Find t when $T = 30$.

(c) Describe what happens to T as t becomes large.

Q8 Solve the equation $2^t + 4 \times 2^{-t} = 5$.

Answers can be found on pages 136–137.

Key points to remember

• **Transformations**

1

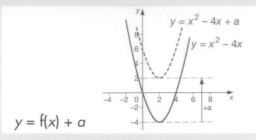

$y = f(x) + a$

A translation of $y = f(x)$ through a units, parallel to the y-axis

2

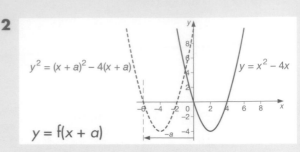

$y = f(x + a)$

A translation of $y = f(x)$ through $-a$ units, parallel to the x-axis

3

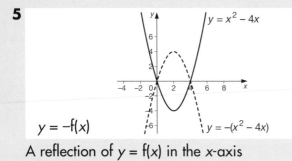

$y = af(x)$

A stretch of $y = f(x)$, scale factor a, parallel to the y-axis

4

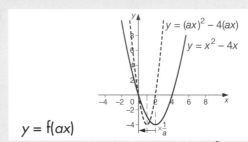

$y = f(ax)$

A stretch of $y = f(x)$, scale factor $\frac{1}{a}$, parallel to the x-axis

The special cases when $a = -1$ in 3 and 4 lead to two more important geometrical transformations:

5

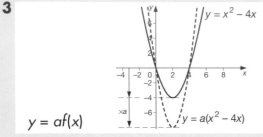

$y = -f(x)$

A reflection of $y = f(x)$ in the x-axis

6

$y = f(-x)$

A reflection of $y = f(x)$ in the y-axis

• An **even function** $f(x)$ has the property $f(-x) = f(x)$.

• An **odd function** $f(x)$ has the property $f(-x) = -f(x)$.

Formulae you must know

• $f(x + a)$ translates $f(x)$ to the left by a units
• $f(x - a)$ translates $f(x)$ to the right by a units
• $f(x) + a$ translates $f(x)$ up by a units
• $af(x)$ stretches $f(x)$ vertically by a factor a
• $f(ax)$ stretches $f(x)$ horizontally by a factor $\frac{1}{a}$

Don't make these mistakes ...

Don't translate the graph the wrong way.

Q1 The graph shows $y = f(x)$.
Sketch the graphs of $y = f(x - 2)$ and $y = f(x) - 2$.

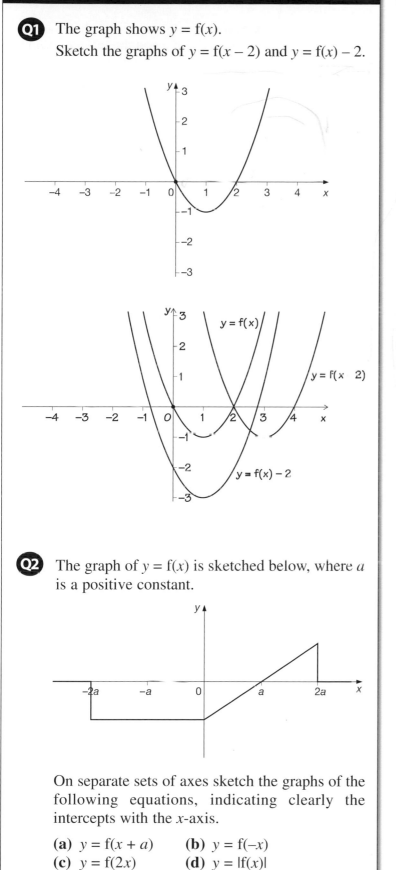

- $y = f(x - 2)$ will be a **translation of 2 units to the right** and $y = f(x) - 2$ will be a **translation of 2 units downwards**. These are shown on the graph.

Q2 The graph of $y = f(x)$ is sketched below, where a is a positive constant.

On separate sets of axes sketch the graphs of the following equations, indicating clearly the intercepts with the x-axis.

(a) $y = f(x + a)$ **(b)** $y = f(-x)$
(c) $y = f(2x)$ **(d)** $y = |f(x)|$

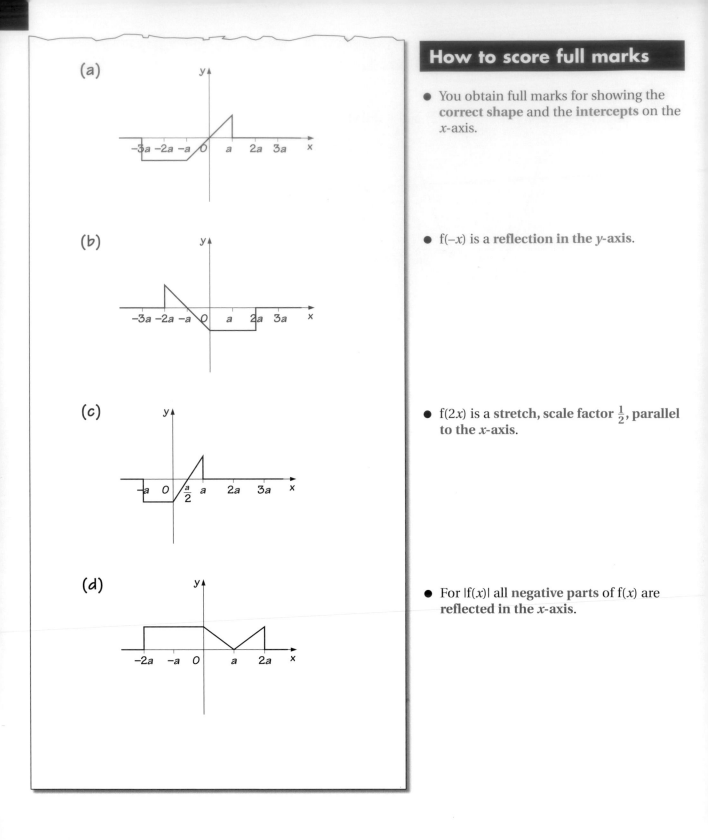

(a)

(b)

(c)

(d)

- You obtain full marks for showing the **correct shape** and the **intercepts** on the *x*-axis.

- f(−*x*) is a **reflection in the *y*-axis**.

- f(2*x*) is a **stretch, scale factor** $\frac{1}{2}$, **parallel to the *x*-axis**.

- For |f(*x*)| all **negative parts** of f(*x*) are **reflected in the *x*-axis**.

Q1 The diagram below shows the graph of $y = x^2$ and three transformations of this graph. Write down the equation of each of the graphs.

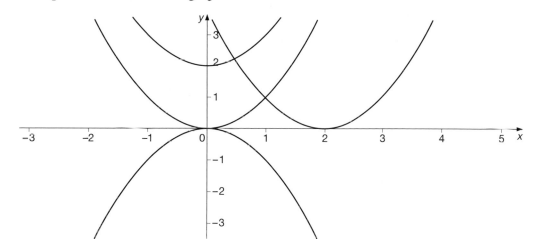

Q2 The figure shows a sketch of the curve with equation $y = f(x)$.

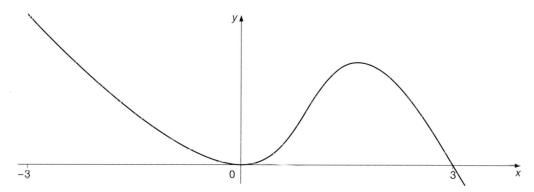

In separate diagrams show, for $-3 \leqslant x \leqslant 3$ a sketch of the curve with equation:

(a) $y = f(-x)$
(b) $y = -f(x)$

marking on each sketch the x-coordinates of any point, or points, where a curve touches or crosses the x-axis.

Q3 The graph below shows the graph of $y = f(x)$.

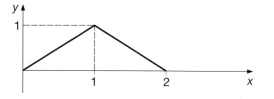

Write down the equations of the following graphs.

(a)

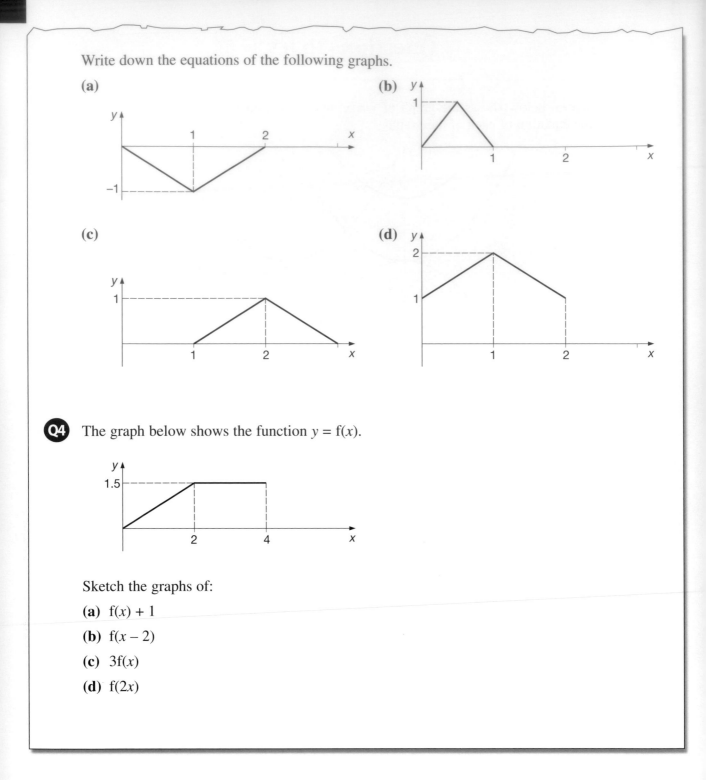

(b)

(c)

(d)

Q4 The graph below shows the function $y = f(x)$.

Sketch the graphs of:

(a) $f(x) + 1$

(b) $f(x - 2)$

(c) $3f(x)$

(d) $f(2x)$

Answers can be found on pages 137–138.

Key points to remember

Constant acceleration formulae

- You will need to know the four formulae and remember that they can only be applied when the acceleration is constant.

Graphs

- The **displacement** of a particle is given by the area enclosed by a velocity-time graph.

- The **acceleration** is given by the gradient of a velocity-time graph.

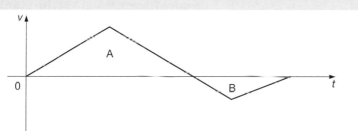

- On this graph:

  the area of triangle A gives the displacement in the **positive direction**

  the area of triangle B gives the displacement in the **negative direction**

  total displacement = area of A − area of B

- The **gradient** of a displacement-time graph gives the **velocity**.

Formulae you must know

- $v = u + at$
- $s = ut + \frac{1}{2}at^2$
- $v^2 = u^2 + 2as$
- $s = \frac{1}{2}(u + v)t$

Don't make these mistakes...

Don't use constant acceleration equations when the **acceleration is not constant**.

Don't forget to consider the **direction of motion**, when finding displacements from a velocity-time graph.

Exam Questions and Student's Answers

How to score full marks

Q1 A lift rises from rest, accelerating at $0.2\,\mathrm{m\,s^{-2}}$ for 2 seconds, then travels at a constant speed for 5 seconds and slows down over a 3-second period.

Find the total distance travelled by the lift.

The speed reached by the lift after 2 seconds is:

$v = 0 + 0.2 \times 2$

$\quad = 0.4\,\mathrm{m\,s^{-1}}$

From the velocity-time graph:

displacement $= \frac{1}{2}(5 + 10) \times 0.4$

$\qquad\qquad\quad = 3\,\mathrm{m}$

- First, you need to calculate the speed at the end of the first 2 seconds using:

$$v = u + at$$

- It is a good idea to draw a **sketch graph** in questions like this. Then you can find the **distance** by calculating the **area under (or enclosed by) the graph**.

- You can find the area by using the **formula for the area of a trapezium**.

$$A = \frac{1}{2}(a + b)h$$

- Alternatively you could split the area into a rectangle and two triangles.

Q2 A ball is thrown vertically upwards, from a height of $1.2\,\mathrm{m}$, with an initial velocity of $4\,\mathrm{m\,s^{-1}}$. Assume that no resistance forces act on the ball.
  (a) Find the maximum height of the ball above the ground.
  (b) Find the speed of the ball when it hits the ground.

(a) Assume the upward direction is positive. At its maximum height $v = 0$ so:

$0^2 = 4^2 + 2 \times (-9.8)s$

$s = \dfrac{4^2}{2 \times 9.8} = 0.82\,\mathrm{m}$

The maximum height above the ground

$= 0.82 + 1.2$

$= 2.02\,\mathrm{m}$

(b) The ball hits the ground when $s = -1.2$.

$v^2 = 4^2 + 2 \times (-9.8) \times (-1.2)$

$v = 6.29\,\mathrm{m\,s^{-1}}$

- In questions like this, you should **define the positive direction**. In this case, the student has defined it as 'upwards'.

- You can take the acceleration (due to gravity) as $-9.8\,\mathrm{m\,s^{-2}}$ and use the initial velocity, which is stated in the question as $4\,\mathrm{m\,s^{-1}}$.

- Now use these values in the formula:

$$v^2 = u^2 + 2as$$

- Remember that the ball was not launched at ground level, so you must **add on the extra height**.

- When the ball hits the ground it will be $1.2\,\mathrm{m}$ below its release point, so s will be -1.2.

Q3 The graph below is a velocity-time graph for a train.

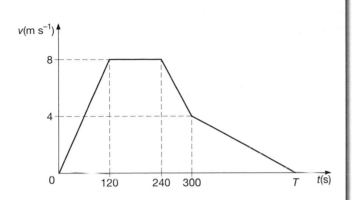

(a) Find the total distance travelled by the train in the first 300 seconds.

(b) Find T if the train travels a total distance of 2 km.

(a) Taking each stage of the motion:

for $t = 0$ to 120,
$$\text{distance} = \frac{1}{2} \times 120 \times 8 = 480 \text{ m}$$

for $t = 120$ to 240,
$$\text{distance} = 120 \times 8 = 960 \text{ m}$$

for $t = 240$ to 300,
$$\text{distance} = \frac{1}{2} \times (8 + 4) \times 60 = 360 \text{ m}$$

$$\text{Total distance} = 480 + 960 + 360$$
$$= 1800 \text{ m}$$

(b) The distance travelled in the final stage must be $2000 - 1800 = 200$ m.

$$200 = \frac{1}{2} \times (T - 300) \times 4$$

$$200 = 2T - 600$$

$$T = 400$$

● Remember that for a **velocity-time graph**, the **distance covered is the area under the graph**.

● The **total area** is made up of a triangle, a rectangle and a trapezium.

● You can use the formulae for each of these areas to find the total distance travelled.

● The **distance travelled on the final stage** is given by the **area of the triangle**. You can use the result from part (a) to find the actual distance.

● Then you can calculate the length of the base of the triangle as $T - 300$.

Q1 A ball is thrown upwards at a speed of $6\,\text{m s}^{-1}$, from a height of $5\,\text{m}$.

(a) Find the maximum height of the ball.   (b) Find the time that the ball is in the air.
(c) Find the speed of the ball when it hits the ground.

Q2 A car starts from rest at a point O and moves in a straight line. The car moves with a constant acceleration $4\,\text{m s}^{-2}$ until it passes the point A, when it is moving with speed $10\,\text{m s}^{-1}$. It then moves with constant acceleration $3\,\text{m s}^{-2}$ for $6\,\text{s}$ until it reaches the point B. Find:

(a) the speed of the car at B   (b) the distance OB.

Q3 A train T_1 moves from rest at station A with constant acceleration $2\,\text{m s}^{-2}$ until it reaches a speed of $36\,\text{m s}^{-1}$. It maintains this constant speed for $90\,\text{s}$ before the brakes are applied, which produces a constant retardation of $3\,\text{m s}^{-2}$. The train T_1 comes to rest at station B.

(a) Sketch a speed-time graph to illustrate the journey of T_1 from A to B.
(b) Show that the distance between A and B is $3780\,\text{m}$.

A second train T_2 takes $150\,\text{s}$ to move from rest at A to rest at B. The diagram shows the speed-time graph illustrating this journey.

(c) Explain briefly one way in which T_1's journey differs from T_2's journey.
(d) Find the greatest speed, in m s^{-1}, attained by T_2 during its journey.

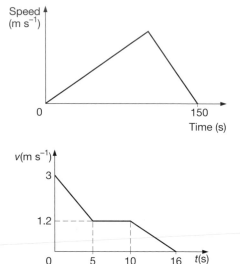

Q4 The graph shows how the velocity of a toy train varies on one section of a track.

(a) Describe briefly the movement of the train.
(b) Find the total distance travelled by the train.
(c) Find the distance travelled by the train when $t = 12$.

Q5 A lift travels from rest from the ground floor and comes to rest again at a height of 20 metres above the ground floor. The motion of the lift takes place in three stages. In the first stage the lift moves with a constant acceleration, it then moves with a constant velocity of $4\,\text{m s}^{-1}$ and finally it moves with a constant retardation until it comes to rest. The times for the three stages of the motion are 2, t and 1 seconds, respectively.

(a) Sketch a velocity-time graph to show the motion of the lift.
(b) Hence, or otherwise, calculate the time for which the lift is in motion.
(c) Calculate the average velocity of the lift during the motion, giving your answer correct to three significant figures.

Q6 A car, travelling at $20\,\text{m s}^{-1}$, passes a point, A. The car moves from A with a constant acceleration of $2\,\text{m s}^{-2}$ until it reaches the point B. It then moves from B to C at a constant speed in a time of 10 seconds. The car travels a total of $425\,\text{m}$.

(a) Find the time that the car takes to travel from A to B.
(b) Find the speed of the car between B and C.

Answers can be found on pages 138–139.

13 Kinematics and vectors

Key points to remember

- When an object such as a boat moves in a current, the resultant velocity is the sum of its own velocity in still water and the velocity of the current.

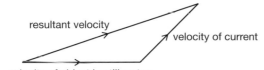

resultant velocity

velocity of current

velocity of object in still water

You can solve problems of this type by scale drawing, but it is generally better to use trigonometry.

- The constant acceleration equations can be used in two dimensions (directions at right angles to each other), where the velocity, acceleration and position are expressed as vectors.

- When an object moves with a constant velocity $\mathbf{v}$ from an initial position $\mathbf{r}_0$, then at time t its position will be given by $\mathbf{v}t + \mathbf{r}_0$.

- Velocities expressed with bearings can be written in terms of the unit vectors $\mathbf{i}$ and $\mathbf{j}$. For example, a velocity of $40\,\text{m s}^{-1}$ on bearing of $070°$ can be written as $40\sin 70°\mathbf{i} + 40\cos 70°\mathbf{j}$.

- Average velocity is the displacement divided by the time taken. This may not be the same as the actual velocity. For example, if an object completes a circle in 20 seconds, the displacement will be zero because the motion begins and ends at the same point. The average velocity will be zero, but clearly the actual velocity will not have been zero.

Formulae you must know

- $\mathbf{v} = \mathbf{u} + \mathbf{a}t$
- $\mathbf{r} = \mathbf{u}t + \frac{1}{2}\,\mathbf{a}t^2$
- $\mathbf{r} = \frac{1}{2}\,(\mathbf{u} + \mathbf{v})t$
- average velocity $= \dfrac{\text{displacement}}{\text{time}}$

Don't make these mistakes...

Don't forget to express quantities as **vectors**.

Don't forget to take account of **initial positions**.

Don't omit $\mathbf{i}$ or $\mathbf{j}$ from expressions.

Q1 Two boats, A and B, set out from the same port at the same time. A travels at a speed of $8\,\text{m s}^{-1}$ on a bearing of 330° and B travels at $5\,\text{m s}^{-1}$ due west. The port is taken to be the origin.

(a) Find expressions for the positions of each boat, relative to their starting points, using the unit vectors **i** and **j**, which are directed east and north respectively.

(b) Determine when the distance between the two boats is equal to 1.4 km.

(a) Velocity of A = $\underline{v}_A = -8\sin 30°\underline{i} + 8\cos 30°\underline{j}$

$\qquad\qquad\quad = -4\underline{i} + 8\cos 30°\underline{j}$

$\underline{r}_A = -4t\underline{i} + 8\cos 30°t\underline{j}$

$\underline{v}_B = -5\underline{i}$

$\underline{r}_B = -5t\underline{i}$

(b) Distance between the boats

$\quad = \sqrt{(-4t - (-5t))^2 + (8\cos 30°t)^2}$

$\quad = \sqrt{t^2 + 48t^2} = \sqrt{49t^2}$

$\quad = 7t$

When $d = 1400$, $1400 = 7t$,

so $t = 200$ seconds.

- Start by expressing the **velocities as vectors**. You should **leave the trigonometric terms in the expressions**, so that you don't introduce any rounding errors.

- The positions of the boats are given by $\mathbf{v}t + \mathbf{r}_0$ where $\mathbf{v}$ is the velocity and $\mathbf{r}_0$ is the initial position.

- To find the distance between the two boats, you need to calculate the **magnitude of the vector $\mathbf{r}_A - \mathbf{r}_B$**.

- Remember that the **velocities are given in m s^{-1}**, so don't forget to **change** the distance of 1.4 km to metres.

Q2 A jet-ski starts from the origin with a velocity of $(4\mathbf{i} + 6\mathbf{j})\,\text{m s}^{-1}$. It accelerates at $(-\mathbf{i} + 2\mathbf{j})\,\text{m s}^{-2}$, for 5 seconds. The unit vectors **i** and **j** are directed east and north respectively.

(a) Find the velocity and speed of the jet-ski after 5 seconds.

(b) Find the distance between the initial and final positions of the jet-ski.

(c) Find the average velocity of the jet-ski over the 5-second period.

(d) Find the time at which the jet-ski is heading due north.

(a) $\underline{v} = (4\underline{i} + 6\underline{j}) + (-\underline{i} + 2\underline{j}) \times 5$

$\quad = -\underline{i} + 16\underline{j}$ m s^{-1}

$\quad v = \sqrt{1^2 + 16^2} = 16.03$ m s^{-1} (4 s.f.)

(b) $\underline{r} = (4\underline{i} + 6\underline{j}) \times 5 + \frac{1}{2} \times (-\underline{i} + 2\underline{j}) \times 5^2$

$\quad = 7.5\underline{i} + 55\underline{j}$

$\quad d = \sqrt{7.5^2 + 55^2} = 55.5$ m (3 s.f.)

(c) Average velocity $= \dfrac{7.5\underline{i} + 55\underline{j}}{5}$

$\quad\quad\quad\quad\quad = 1.5\underline{i} + 11\underline{j}$

(d) $\underline{v} = (4\underline{i} + 6\underline{j}) + (-\underline{i} + 2\underline{j})t$

$\quad = (4 - t)\underline{i} + (6 + 2t)\underline{j}$

If the jet-ski is travelling due north, the velocity in the east-direction must be 0.

This means $4 - t = 0$

$\Rightarrow t = 4$

Q3 An aeroplane flies due north at a speed of 80 m s^{-1} relative to the air. It flies through air that is moving north-east at 30 m s^{-1}. Find the resultant velocity of the aeroplane, expressing it as a speed and the bearing on which the aeroplane is actually moving.

$v^2 = 30^2 + 80^2 - 2 \times 30 \times 80 \times \cos 135°$

$v = 103$ m s^{-1} (3 s.f.)

The bearing will be α.

$\dfrac{\sin\alpha}{30} = \dfrac{\sin 135°}{103}$

$\alpha = 12°$ to the nearest degree

The aeroplane is travelling at 103 m s^{-1} on a bearing of 012°.

How to score full marks

- Use the motion equation $\mathbf{v} = \mathbf{u} + \mathbf{a}t$ to find the velocity.
- Remember that the speed is the **magnitude** of the velocity.
- Now you can find the final position, using $\mathbf{r} = \mathbf{u}t + \frac{1}{2}\mathbf{a}t^2$, then the magnitude of $\mathbf{r}$ gives the **distance**.

- The **average velocity** is the displacement, found in (b), divided by the time.

- The jet-ski will be heading north when the **i**-component is 0. You need to know this, then you can form the equation and solve it.

- By drawing a clear sketch, you can show the examiner that you understand the question and avoid making errors.
- Remember that the speed is represented by the length (or **magnitude**) of the **resultant velocity vector**.
- You need to remember both the **cosine rule** and the **sine rule** for this type of problem. In this example, you use the cosine rule first, to find the **speed** (a length), and then the sine rule to find the **bearing** (an angle).

Questions to try

Q1 During a 10-second period the velocity of a boat changes from $(4\mathbf{i} + 2\mathbf{j})\,\text{m s}^{-1}$ to $(\mathbf{i} - 3\mathbf{j})\,\text{m s}^{-1}$, where $\mathbf{i}$ and $\mathbf{j}$ are perpendicular unit vectors. Find the acceleration of the boat during this time, assuming that it is constant.

Q2 An object has initial velocity $(5\mathbf{i} - 5\mathbf{j})\,\text{m s}^{-1}$ and an acceleration of $(-\mathbf{i} + 2\mathbf{j})\,\text{m s}^{-2}$, where $\mathbf{i}$ and $\mathbf{j}$ are unit vectors directed east and north respectively.

  (a) If the object starts at the origin, find expressions for its position and velocity after t seconds.
  (b) Find the time at which the object is travelling due north.

Q3 Two model boats on a pond are set into motion. The unit vectors $\mathbf{i}$ and $\mathbf{j}$ are east and north respectively. Boat A has a constant velocity of $(4\mathbf{i} + 3\mathbf{j})\,\text{m s}^{-1}$ and starts at a point with position vector $4\mathbf{j}$. Boat B has a constant velocity of $(2\mathbf{i} - \mathbf{j})\,\text{m s}^{-1}$ and initial position $(4\mathbf{i} + 12\mathbf{j})\,\text{m}$.

  (a) Find the position vector of each boat at time t seconds.
  (b) Show that the boats collide and find the position of the boats at this time.
  (c) Find the distance between the two boats when $t = 1$.

Q4 A swimmer can move through still water at $1.2\,\text{m s}^{-1}$. She swims in a straight line in a river flowing at $0.8\,\text{m s}^{-1}$. She travels from the point A to the point B, so that her resultant velocity makes an angle of $30°$ to the downstream bank, as shown in the diagram.

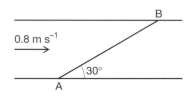

  (a) Sketch an appropriate triangle of velocities.
  (b) Use either scale drawing or trigonometry to find the magnitude of her resultant velocity.

Q5 At time $t = 0$, a boat is at the origin travelling due east with speed $3\,\text{m s}^{-1}$. The boat experiences a constant acceleration of $(-0.2\mathbf{i} - 0.3\mathbf{j})\,\text{m s}^{-2}$. The unit vectors $\mathbf{i}$ and $\mathbf{j}$ are directed east and north respectively.

  (a) Write down the initial velocity of the boat.
  (b) Find an expression for the position of the boat at time t seconds.
  (c) Find the time when the boat is due south of the origin.
  (d) Find the distance of the boat from the origin when it is travelling south east.

Q6 Two cars A and B are moving on straight, horizontal roads with constant velocities. The velocity of A is $20\,\text{m s}^{-1}$ due east, and the velocity of B is $(10\mathbf{i} + 10\mathbf{j})\,\text{m s}^{-1}$, where $\mathbf{i}$ and $\mathbf{j}$ are unit vectors directed due east and due north respectively. Initially A is at the fixed origin O, and the position vector of B is $300\mathbf{i}\,\text{m}$ relative to O. At time t seconds, the position vectors of A and B are $\mathbf{r}$ and $\mathbf{s}$ metres respectively.

  (a) Find expressions for $\mathbf{r}$ and $\mathbf{s}$ in terms of t.
  (b) Hence write down an expression for $\overrightarrow{AB}$ in terms of t.
  (c) Find the time when the bearing of B from A is $045°$.
  (d) Find the time when the cars are $300\,\text{m}$ apart again.

Answers can be found on pages 139–140.

Key points to remember

- Identify the **different forces acting on a particle** and draw clear force diagrams showing these forces.
- The common forces that occur in most problems are **gravity, friction, normal reaction, tension in a string** and the **tension or compression in a rod**.
- In **equilibrium** a particle is at rest or moving with constant velocity. There is **no change in motion**.
- In **equilibrium** the **resultant force is zero** and the **resultant moment is zero**. If the particle is not at rest or moving with constant velocity then Newton's second law is the equation of motion.
- For problems involving connected particles, apply **Newton's second law to each particle** in the system separately.

Formulae you must know

- The law of friction: $F \leqslant \mu R$
  where F is the friction, μ is the coefficient of friction and R is the normal reaction
- Newton's second law: resultant force = mass $\times$ acceleration
- For a particle to remain at rest, the resultant force must be zero
- 1 tonne $= 1000\,\text{kg}$

Don't make these mistakes...

Don't always assume that the **normal reaction** is equal to the weight.

Take care when **resolving** – don't use the wrong angle.

When drawing or labelling a **force diagram**, only include **forces that actually exist**.

Don't forget to include **negative signs** for forces that act to **oppose the motion**.

Exam Questions and Student's Answers

How to score full marks

Q1

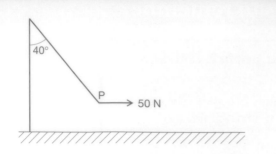

A bowling ball P is attached to one end of a light, inextensible string, the other end of the string being attached to the top of a fixed vertical pole. A girl applies a horizontal force of magnitude 50 N to P, and P is in equilibrium under gravity with the string making an angle of 40° with the pole, as shown in the diagram.

By modelling the ball as a particle, find, to three significant figures,

(a) the tension in the string
(b) the weight of P.

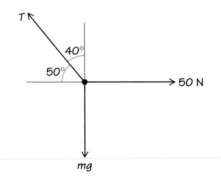

- You must always start with a clear **force diagram** showing the forces on the ball. This shows the examiner that you understand the question, and helps you to avoid making mistakes.

(a) Resolving the forces horizontally:

$T\cos 50° = 50$

$T = 77.8$ N

(b) Resolving the forces vertically:

$T\cos 40° = mg$ (the weight)

$mg = 59.6$ N

The weight of the ball is 59.6 N.

- Since the bowling ball is held **at rest** the components of the forces must **balance in any direction**.
- In this example, it is sensible to resolve the forces into horizontal and vertical components.

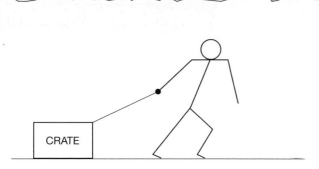

Q2

The diagram shows a crate of weight 300 newtons which is being pulled along a rough horizontal surface at constant speed. The tension in the rope pulling the crate is 100 newtons, and the rope makes an angle of 20° with the horizontal.

(a) Draw a diagram showing the forces acting on the crate.
(b) Calculate the value of:
    (i) the normal reaction force
    (ii) the friction force.
(c) Calculate the value of the coefficient of friction between the crate and the surface.

(a)

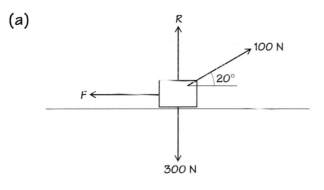

(b) (i) Resolving the forces vertically:

$$R + 100\sin 20° = 300$$

$$R = 300 - 100\sin 20°$$

$$R = 266 \text{ N (to 3 s.f.)}$$

(ii) Resolving the forces horizontally:

$$F = 100\cos 20° = 94.0 \text{ N (to 3 s.f.)}$$

(c) Applying the law of friction:

$$F = \mu R \Rightarrow \mu = \frac{100\cos 20°}{300 - 100\sin 20°} = 0.354$$
$$\text{(to 3 s.f.)}$$

- The forces are the normal reaction R, friction force F, the weight and the tension in the string.

- The crate is moving with constant speed along the surface, so the forces are in balance.
- Use $R + 100\sin 20° = 300$
  or $R + 100\cos 70° = 300$.

- Remember that, **since the crate is sliding**, the friction law is:

$$F = \mu R$$

Q3 Two particles are connected by a light inelastic string that passes over a smooth, light pulley. The particles have masses of 9 kg and 5 kg. They are released from rest with the strings vertical.

(a) By using an equation of motion for each particle, find the acceleration of the masses.
(b) Find the tension in the string.

(a) For 9 kg mass

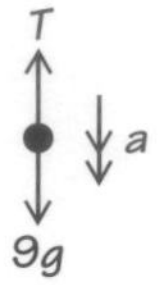

$$T \uparrow \quad \downarrow a$$
$$9g$$

Resultant Force = $9g - T$

$9g - T = 9a$

For 5 kg mass

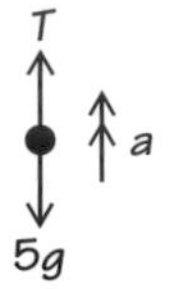

$$T \uparrow \quad \uparrow a$$
$$5g$$

Resultant Force = $T - 5g$

$T - 5g = 5a$

From equation (2)

$T = 5a + 5g$

Substituting into equation (1)

$9g - (5a + 5g) = 9a$

$9g - 5a - 5g = 9a$

$4g = 14a$

$a = \dfrac{4g}{14} = 2.8 \text{ ms}^{-2}$

(b) $T = 5a + 5g$

$T = 5 \times 2.8 + 5 \times 9.8 = 63 \text{ N}$

How to score full marks

- It is important that you **consider each particle in turn**. Also note that the **tension will be the same throughout the string**.

- For each particle find the resultant force and then apply **Newton's second law** ($F = ma$).

- When you have the two equations, one for each particle, solve them to find a.

- To find the tension, go back to one of your earlier equations and substitute the value of a.

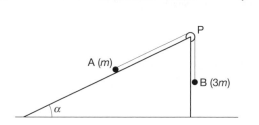

The particle, A, of mass m rests on a rough plane inclined at an angle α to the horizontal, where $\tan \alpha = \frac{3}{4}$.

The particle is attached to one end of a light, inextensible string which passes over a small, light smooth pulley P fixed at the top of the plane. The other end of the string is attached to a particle B of mass $3m$, and B hangs freely below P. The particles are released from rest with the string taut. The particle B moves down with acceleration of magnitude $\frac{1}{2}g$. Find:

(a) the tension in the string

(b) the coefficient of friction between A and the plane.

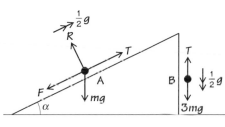

$\tan \alpha = \frac{3}{4} \qquad \cos \alpha = \frac{4}{5} \qquad \sin \alpha = \frac{3}{5}$

(a) Applying Newton's second law to B:

$$3mg - T = 3m \times \frac{1}{2}g \Rightarrow T = \frac{3}{2}mg$$

(b) Apply Newton's second law to A.

Perpendicular to plane:
$$R - mg\cos \alpha = 0 \Rightarrow R = \frac{4}{5}mg$$

Parallel to plane:
$$T - mg\sin \alpha - F = m \times \frac{1}{2}g$$

$$F = T - mg\sin \alpha - \frac{1}{2}mg$$

$$F = \frac{3}{2}mg - \frac{3}{5}mg - \frac{1}{2}mg = \frac{2}{5}mg$$

The law of sliding friction gives $F = \mu R$.

Coefficient of friction $\mu = \dfrac{F}{R} = \dfrac{\frac{2}{5}mg}{\frac{4}{5}mg} = \dfrac{1}{2}$

- Start by writing the forces on the diagram. You might also find it helpful to write down useful information, such as the trigonometric ratios, as this student has done.

- **Newton's second law** is the equation of motion.

- Remember that **friction F opposes the direction of motion** of A.

- Remember that $F \leqslant \mu R$ if the object is at rest. Otherwise $F = \mu R$ when the object is **sliding or about to slide**.

Q1 A sledge is modelled as a particle of mass 15 kg. The diagram shows the forces that act on the sledge as it is pulled across a rough, horizontal surface. The coefficient of friction between the sledge and the ground is 0.4.

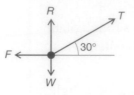

(a) Show that the weight, W, of the sledge is 147 N.

(b) Given that $T = 80$ N, show that $R = 107$ N.

(c) Find the magnitude of the friction force acting on the sledge.

(d) Find the acceleration of the sledge.

(e) The sledge is initially at rest. Find the speed of the sledge after it has been moving for 3 seconds.

Q2

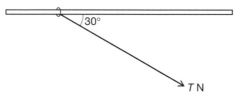

A heavy ring of mass 5 kg is threaded on a fixed rough horizontal rod. The coefficient of friction between the ring and the rod is $\frac{1}{2}$. A light string is attached to the ring and is pulled with a force of magnitude T newtons acting at an angle of $30°$ to the horizontal (see diagram). Given that the ring is about to slip along the rod, find the value of T.

Q3 A block, of mass 20 kg, rests on a rough horizontal surface. A light, inextensible string attached to the block passes over a smooth, light pulley. A weight, of mass 10 kg, hangs from the other end of the string, as shown in the diagram below.

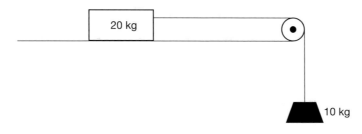

When released from rest the system accelerates at 0.5 m s^{-2}.
Find the coefficient of friction between the block and the surface.

Q4 A sign is hung outside a shop. It has mass m kg and is held in equilibrium by two strings. The sign is modelled as a particle P. One string is inclined at 50° to the vertical and exerts a force of 60 N on the particle. The other string exerts a force of magnitude T N at an angle of 48° to the vertical. The forces that act on the particle are shown in the diagram below.

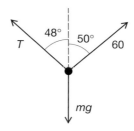

(a) Find T.

(b) Find m.

Q5 Two particles are connected by a light, inextensible string which passes over a smooth, light pulley. The particles are of mass 2 kg and 3 kg. The system is shown in the diagram below.

The system is released from rest with the string taut.

(a) By forming an equation of motion for each particle, show that the tension in the string is 23.52 N.

(b) Find the magnitude of the acceleration of the particles.

Answers can be found on pages 141–142.

15 Conservation of momentum

Key points to remember

- **Momentum** is the product of the **mass and velocity** of a body.
- In collisions, **momentum is conserved** if no external forces act.
- The change in momentum of a body is the **impulse**.*
- When a **constant force** acts, the impulse is equal to the product of the force and the time for which it acts.*

Formulae you must know

- $m_A u_A + m_B u_B = m_A v_A + m_B v_B$
- $I = mv - mu$*
- $I = Ft$*

* Not needed for all examination boards.

Don't make these mistakes...

Don't forget to **include the signs** when working with negative velocities.

Don't forget to define a positive direction, when starting a problem.

Exam Questions and Student's Answers

Q1 Two particles are moving towards each other. Particle A has mass 2 kg and speed 3 m s^{-1}. Particle B has mass 4 kg and speed 5 m s^{-1}. After the collision particle B moves in the same direction, but its speed has been reduced to 2 m s^{-1}. Describe the motion of particle A after the collision.

Using conservation of momentum:

$$2 \times 3 + 4 \times (-5) = 2v_A + 4 \times (-2)$$

$$-14 = 2v_A - 8$$

$$v_A = -3$$

So after the collision particle A has changed direction and is moving at 3 m s^{-1}.

Q2 A train carriage, of mass 50 tonnes, is travelling at 10 m s^{-1}, when it collides with a second carriage, of mass 100 tonnes, travelling at 6 m s^{-1} in the same direction. After the collision the two carriages travel together at the same speed.

Find the speed of the carriages after the collision.

Using conservation of momentum:

$$50\,000 \times 10 + 100\,000 \times 6 = 150\,000v$$

$$1\,100\,000 = 150\,000v$$

$$v = 7.33 \text{ (3 s.f.)}$$

The speed after the collision is 7.33 m s^{-1}.

How to score full marks

- The student has drawn a diagram to show the velocities of the particles **before and after the collision**. This is a very good way to start problems like this one, as it gives a summary of the information you need.

  Note that:

  $m_A = 2 \quad u_A = 3 \quad v_B = -2$

  $m_B = 4 \quad u_B = -5$

- You can substitute these values into the **equation for the conservation of momentum**.

- Again, the student has started with a diagram to show the information given in the question.
- The diagram shows the masses and velocities of the carriages **before and after the collision**.
- Note that after the collision the two carriages can be considered as a **single particle**. In this case:

  $m_A = 50\,000 \quad u_A = 10 \quad v_A = v_B = v$

  $m_B = 100\,000 \quad u_B = 6$

- If you make a list of these when you start, then you can substitute them into the **conservation of momentum equation**.

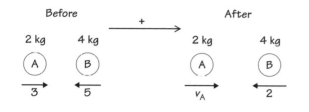

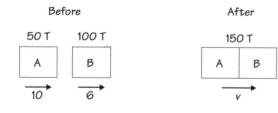

Q1 Two particles A and B have masses of 5 kg and 3 kg respectively and move on a straight line. They collide and coalesce during the collision, moving as a single particle after the collision. Before the collision, the speed of A is 6 m s^{-1} and speed of B is 4 m s^{-1}.

Find the speed of the combined particle if A and B were moving in:

(a) the same direction before the collision

(b) opposite directions before the collision.

Q2 A child, of mass 48 kg, jumps off a stationary skateboard. He initially moves horizontally at 0.5 m s^{-1}. The skateboard has mass 1.2 kg. Calculate the initial speed of the skateboard.

Q3

2 m s^{-1} 3 m s^{-1}

x kg   0.1 kg

Two particles, of masses x kg and 0.1 kg, are moving towards each other in the same straight line and collide directly. Immediately before the impact, the speeds of the particles are 2 m s^{-1} and 3 m s^{-1} respectively (see diagram).

(a) Given that both particles are brought to rest by the collision, find x.

(b) Given instead that the particles move with equal speeds of 1 m s^{-1} after the impact, find the three possible values of x.

Q4 Three particles A, B and C lie in a straight line on a smooth horizontal surface. The masses of the three particles are 2 kg, 3 kg and m kg respectively.

(a) The particle A is set into motion, so that it moves towards B with speed 6 m s^{-1}. When it collides with B the two particles coalesce and move with speed v m s^{-1} towards C. Find v.

(b) The combined particle then hits C. After this collision, C moves with a speed of 0.7 m s^{-1} and the combined particle (A and B) travels with a speed of 0.4 m s^{-1} in the opposite direction. Find m.

Answers can be found on page 142.

Key points to remember

- The **position** of a projectile launched from ground level is given by $x = V\cos\theta t$ and $y = V\sin\theta t - \frac{1}{2}gt^2$.

- The **position** of a projectile launched **from a height** **h** is given by $x = V\cos\theta t$ and $y = V\sin\theta t - \frac{1}{2}gt^2 + h$.

- To find the **time of flight** of a projectile solve the equation $V\sin\theta t - \frac{1}{2}gt^2 = 0$.

- To calculate the **range** of a projectile substitute the time of flight into the equation $x = V\cos\theta t$.

- At its **maximum height** the vertical component of the velocity of the projectile will be zero.

- These are the key modelling assumptions for the motion of a projectile:
  - (i) weight or force of gravity is the only force that acts on the projectile
  - (ii) there is no air resistance, lift, etc.
  - (iii) projectile is a particle.

Formulae you must know

- $x = V\cos\theta t$

- $y = V\sin\theta t - \frac{1}{2}gt^2$

Don't make these mistakes...

Don't ignore the height at which a projectile is launched.

Don't rely on formulae that you have learned for the range, time of flight and maximum height. You should derive these, as you need them. Students make errors when quoting the formulae and often the questions are set so that these formulae do not apply.

Don't break the motion of the projectile up into a lot of stages. Normally one equation can be used which will apply to the whole motion of the projectile.

Q1 A football is kicked from horizontal ground. It initially has a velocity of 20 m s⁻¹ at an angle of 60° above the horizontal. The football hits the ground T seconds after it was kicked.

(a) Find T.
(b) Find the range of the football.
(c) Find the maximum height of the football.

(a) $0 = 20\sin60°T - 4.9T^2$

$0 = T(20\sin60° - 4.9T)$

$T = 0$ or $20\sin60° - 4.9T = 0$

$T = \dfrac{20\sin60°}{4.9} = 3.53$ seconds

(b) Range $= 20\cos60° \times \dfrac{20\sin60°}{4.9} = 35.3$ m

(c) $t = \dfrac{1}{2} \times \dfrac{20\sin60°}{4.9} = \dfrac{10\sin60°}{4.9}$

$H = 20\sin60° \times \dfrac{10\sin60°}{4.9} - 4.9\left(\dfrac{10\sin60°}{4.9}\right)^2$

$= 15.3$ m

● You first need to form an equation to find the time when the football hits the ground for the first time by using $y = V\sin\theta\, t - \dfrac{1}{2}gt^2$. Solving this gives the time of flight.

● Substitute the time of flight into the equation $x = V\sin\theta\, t$ to find the range.

● As the football starts and finishes at the same level the **maximum height** will be reached **half way through the flight**. This time can be substituted into $y = V\sin\theta\, t - \dfrac{1}{2}gt^2$ to find the maximum height

Q2 A shot putter throws a shot at an initial speed of 12 m s⁻¹, and at an angle of 40° above the horizontal. Assume that the shot is released at a height of 2 m.

(a) Find the range of the shot.
(b) Find the maximum height of the shot.

(a) $-2 = 12\sin40°t - 4.9t^2$

$0 = 4.9t^2 - 12\sin40°t - 2$

$t = \dfrac{12\sin40° \pm \sqrt{(12\sin40°)^2 - 4\times4.9\times(-2)}}{4\times4.9}$

$= -0.2267$ or 1.801

$t = 1.801$ s

$R = 12\cos40° \times 1.801$

$= 16.6$ m

● First find the **time of flight** of the projectile, noting that when it lands it will have a displacement of –2 compared to its initial position.

● Once the time of flight has been found it can be multiplied by the horizontal component of the velocity to determine the **range**.

(b) $0 = 12\sin 40° - 9.8t$

$$t = \frac{12\sin 40°}{9.8}$$

$$H = 12\sin 40° \times \frac{12\sin 40°}{9.8} - 4.9\left(\frac{12\sin 40°}{9.8}\right)^2 + 2$$

$$= 5.04 \text{ m}$$

Q3 A bullet is fired horizontally from a rifle at a speed of 120 m s^{-1} and at a height of 3 metres. Assume that the ground is horizontal and that the bullet is not affected by air resistance.

(a) Find the time that the bullet is moving before it hits the ground.

(b) Find the horizontal distance of the bullet from the rifle.

(a) $y = 3 - 4.9t^2$

$0 = 3 - 4.9t^2$

$$t^2 = \frac{3}{4.9}$$

$$t = \sqrt{\frac{3}{4.9}} = 0.782 \text{ seconds}$$

(b) $x = 120 \times \sqrt{\frac{3}{4.9}} = 93.9 \text{ m}$

How to score full marks

- The **maximum height** will be reached when the **vertical component of the velocity is zero**. Solving this equation gives the time when the projectile is at its maximum height. The actual height can then be calculated using this time.

- Note that the vertical equation does not involve the initial velocity as this does not have a vertical component. In this case you can use $y = h - \frac{1}{2}gt^2$, where h is the initial height.

- Note that the **horizontal displacement** is simply given by $x = 120t$.

Questions to try

Q1 A golf ball is placed on a horizontal surface and hit, so that it initially moves with velocity 40 m s^{-1} at an angle of $35°$ above the horizontal.

(a) Find the time of flight of the ball.

(b) Find the range of the ball.

(c) Find the maximum height of the ball.

Q2 A basketball is thrown from a height of 1 metre. Its initial velocity is 12 m s^{-1} at an angle of $60°$ above the horizontal. The ball passes through a basket at a height of 3 metres.

(a) Find the time that it takes the ball to reach the basket.

(b) Find the horizontal distance from the point where the ball was thrown to the basket.

(c) Find the maximum height of the ball above ground level.

Q3 A cricket ball is hit from horizontal ground with a speed of 26 m s^{-1} at an angle of $20°$ above the horizontal. The motion of the ball is modelled as that of a particle moving freely under gravity.

(a) Find the greatest height above the ground reached by the ball.

(b) The ball has travelled a horizontal distance of 30 m when it hits a wall. Find the height of the ball when it hits the wall.

Q4 An arrow is fired horizontally from a height of 1.5 metres with speed 12 m s^{-1}.

(a) Find the time taken for the arrow to reach the ground.

(b) Find the horizontal distance travelled by the arrow.

(c) Find the speed of the arrow when it hits the ground.

Q5 A ball is hit from a horizontal surface with a speed of 8 m s^{-1}. It hits the ground again 1.2 seconds later.

(a) Find the angle between the initial velocity of the ball and the horizontal.

(b) Find the range of the ball.

Answers can be found on page 143.

Key points to remember

- The most common measure of average is the **mean**. It is usually preferred to other measures because all the data items contribute to the final value.

- For data sets which are highly skew or which contain outliers the mean may be well away from the centre of the data set and so the **median** may be preferred.

- The **mode** is the most commonly occurring value but is of little use if the data is sparse. For continuous variables the **modal class** can be used.

- The mean of a population is usually denoted by μ.

- The mean of a sample (part of the population) is usually denoted by $\bar{x}$.

- The most common measure of spread is the **standard deviation**. It is associated with the mean and has the same advantages and disadvantages.

- The **interquartile range** is associated with the median and the **range** is associated with the mode.

- The standard deviation of a population is denoted by σ.

- When (as is nearly always the case) the population standard deviation is estimated from a sample, s is used as an estimate of σ.

- The **variance** is the standard deviation squared. It has a number of properties which make it useful in mathematical statistics but it is not useful as a measure of spread. This is because it is in the wrong units (e.g. if the data is in Euros the variance is in square Euros).

Don't make these mistakes...

You have probably met all these measures at GCSE. Don't assume that evaluating them is easy and it is unnecessary to revise them. On AS papers the questions on numerical measures are usually poorly answered.

Understanding the difference between σ and s and identifying the exact limits of the classes for grouped data are two of the most difficult items in S1.

Formulae you must know

Some boards, such as AQA, will encourage you to find the mean and standard deviation directly from a calculator but other boards may expect you to show details of your working or only give the data in the form of summations.

- **For a sample x_1, x_2, x_3, ..., x_n**

$$\bar{x} = \frac{\sum x_i}{n} \qquad s = \sqrt{\frac{\sum(x_i - \bar{x})^2}{n-1}} = \sqrt{\frac{\sum x_i^2 - \frac{(\sum x_i)^2}{n}}{n-1}}$$

If the observations x_1, x_2, x_3, ..., x_n occur with frequencies f_1, f_2, f_3, ..., f_n respectively, the formulae become

$$\bar{x} = \frac{\sum f_i x_i}{\sum f_i} \qquad s = \sqrt{\frac{\sum f_i(x_i - \bar{x})^2}{(\sum f_i) - 1}} = \sqrt{\frac{\sum f_i x_i^2 - \frac{(\sum f_i x_i)^2}{\sum f_i}}{(\sum f_i) - 1}}$$

- **For a population x_1, x_2, x_3, ..., x_n**

$$\mu = \frac{\sum x_i}{n} \qquad \sigma = \sqrt{\frac{\sum(x_i - \mu)^2}{n}}$$

Exam Questions and Student's Answers

Q1 A sample of people who commuted daily from a town in Cheshire into Manchester were asked to estimate the time taken on their most recent journey. The replies are summarised below.

Time (minutes)	Frequency
35–	8
45–	14
55–	23
75–	12
95–125	9

(a) Calculate estimates of the mean and standard deviation of these times.

(b) A sample of people who commuted regularly from Merseyside into Manchester were asked the same question. Their answers had a mean of 67.9 minutes with a standard deviation of 11.5 minutes.

(i) Compare the distributions of the two samples.

(ii) Give two reasons why the data may not provide good estimates of commuting times from the two locations.

(a)

Time	Class mid-mark, x	f	fx	fx^2
35–	40	8	320	12800
45–	50	14	700	35000
55–	65	23	1495	97175
75–	85	12	1020	86700
95–125	110	9	990	108900

$\sum f = 66 \quad \sum fx = 4525$

$\sum fx^2 = 340575$

$\bar{x} = 4525/66 = 68.6$

$s = \sqrt{(340575 - 4525^2/66)/(66-1)}$

$= 21.6$

How to score full marks

- The accuracy of measurement is not stated so assume that the class '35–' includes all times from 35 up to 45. Hence the **class mid-mark** is $(35 + 45)/2 = 40$.

- Once the class mid-marks have been identified, the **mean and standard deviation can be found directly from a calculator**. This is the recommended method unless the question requires you to show full details – as opposite.
- Since the data is stated to be a **sample**, use s (divisor 66 – 1) rather than σ (divisor 66).

(b) (i) The average commuting time is similar but the times from Merseyside are much less variable.

(ii) There is no information on how the samples were selected; the times were estimated after the journeys were completed rather than measured; no information on whether there were any unusual delays or weather conditions when the journeys were undertaken, etc.

● **Any two sensible comments** will score full marks.

Q2 It is proposed that applicants for work at a small engineering firm should be given an aptitude test consisting of a mechanical puzzle. The current employees of the firm were timed to complete the puzzle and the times taken by 45 of them are shown below.

Time to complete puzzle (seconds)	Frequency
10–29	7
30–39	11
40–44	9
45–49	8
50–89	10

(a) Draw a cumulative frequency diagram of the data and estimate the median and interquartile range.

(b) Calculate estimates of the mean and standard deviation of the data.

(c) In addition to the data in the table, five other employees attempted the puzzle but after 90 seconds had failed to complete it. Estimate the median time to complete the puzzle for all 50 employees.

(d) Explain why the median might be preferred to the mean as a measure of average in these circumstances.

(a)

Time	Class mid-mark	Upper class bound	Frequency	Cumulative frequency
10–29	19.5	29.5	7	7
30–39	34.5	39.5	11	18
40–44	42.0	44.5	9	27
45–49	47.0	49.5	8	35
50–89	69.5	89.5	10	45

How to score full marks

- There are **gaps between the class bounds**, e.g. the observation 29.7 will not fit into any class. The data must have been recorded to the nearest second. Hence the **real boundaries** of the class '10–29' are 9.5 and 29.5.

- Plot the **cumulative frequency against the upper class bound**.

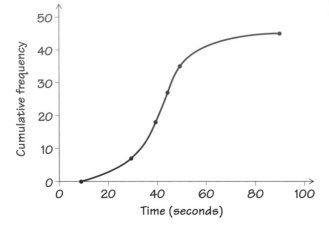

Median = 42 seconds.

Lower quartile = 34, Upper quartile = 49

Interquartile range = 49 − 34 = 15 seconds.

(b) Using the class mid-mark, by calculator

$\bar{x} = 43.7 \qquad s = 16.4$

(c) Although the exact times for five employees are unknown, they are all known to exceed 90 seconds. Hence the median time for all 50 employees may be found by reading off the cumulative frequency curve at (50+1)/2= 25.5

Median = 43.5 seconds.

(d) The median can be found from the given data but the mean cannot be calculated as the times of five employees are unknown. An additional reason is that the median, unlike the mean, would not be affected if a few of the employees took an exceptionally long time to complete the puzzle.

- The **median** is read off from the graph at a cumulative frequency of (45 +1)/2 = 23.

- The **lower quartile** is read off at (45 + 1)/4 = 11.5 and the **upper quartile** at 3(45 + 1)/4 = 34.5.

- These can be calculated as in Q1 but it is quicker and easier to find them direct from a calculator.

Q3 The heights of a sample of female army recruits are summarised in the table below.

Height (cm)	Frequency
131–140	4
141–150	23
151–155	16
156–160	12
161–170	18
171–190	11

(a) For the data above find:
  (i) the mean
  (ii) the modal class.

(b) It is decided to order a set of uniforms to fit recruits of average height. Uniforms for recruits whose height is too far from the average will have to be made individually. State, giving a reason, the most suitable height to order uniforms for.

(a)

Height (cm)	Class mid-mark	Frequency	Frequency density
131–140	135.5	4	0.4
141–150	145.5	23	2.3
151–155	153	16	3.2
156–160	158	12	2.4
161–170	165.5	18	1.8
171–190	180.5	11	0.55

(i) From calculator the mean is 157.1 cm.
(ii) Modal class is 151–155.

(b) This data is skewed and the tall recruits have pulled the mean height away from the centre of the distribution. As we wish to order uniforms to fit as many recruits as possible the mode is the most appropriate measure of average. As we only have a modal class and a single height is required, choose the mid point, i.e. 153 cm.

- The heights must have been recorded to the nearest cm and the actual limits of the class '131–140' are 130.5 and 140.5.
- If the classes were of equal width the class containing the largest frequency would be the modal class. As the classes are not of equal width the **modal class** is the class with the **largest frequency density** (the frequency density is the frequency divided by the class width).
- Don't forget that the **actual boundaries** of the class '131–140' are 130.5 to 140.5 and so the width of the class is 10.

Questions to try

(In these questions assume that the data is a sample and not the whole population.)

Q1 The number of telephone calls made by Ella on 15 successive days was:

6   1   3   4   0   4   2   3   9   4   4   3   5   0   4

Find the:
(i) mean   **(ii)** median   **(iii)** mode   **(iv)** standard deviation   **(v)** interquartile range   **(vi)** range.

Q2 The lengths of the 52 telephone calls made by Ella are summarised in the following table:

Time (seconds)	Frequency
20–	5
80–	17
140–	15
200–	9
260–320	6

Find the:
(i) mean   **(ii)** median   **(iii)** modal class   **(iv)** standard deviation   **(v)** interquartile range.

Q3 The value, £, of the weekly orders taken over an eighteen-week period by Samir, a pharmaceutical company representative, are shown below:

1450 1380 1670 1880 1180 2270 2070 1940 1530 1300 2150 2270 1120 1990 1540 1410 2400 1760

The value, £, of the weekly orders taken over the same eighteen-week period by Marian, a representative for the same company, are shown below:

1340 1590 1570 1780 1080 1670 1970 1740 1430 1600 2050 1860 1530 1880 1450 1510 1800 1660

Calculate the mean and standard deviation for each representative and compare their performance.

Q4 Calculate the median and interquartile range for the data in question 3, and again compare the performance of the two representatives.

Q5 The lifetimes, in hours, of a sample of 100 abrasive discs are summarised in the following table.

Lifetime (hours)	Frequency
650–699	9
700–719	19
720–729	18
730–739	19
740–759	22
760–809	13

Find the:
(i) mean   **(ii)** standard deviation   **(iii)** median   **(iv)** interquartile range   **(v)** modal class.

Q6 A company sells two makes of television set, Rony and Pitachi, and provides free after sales service.

For Rony sets bought in June 2001, the times, in days, from installing a machine to first being called out to deal with a breakdown are summarised in the following table.

Time to first call-out (days)	Frequency
0–	8
100–	22
200–	31
400–	29
800–1200	14

(a) By drawing a cumulative frequency diagram or otherwise estimate the median and the interquartile range.

(b) For Pitachi sets sold in June 2001, the median time from installing a set to first being called out to deal with a breakdown was 458 days and the interquartile range was 660 days. Compare, briefly, the reliability of Rony and Pitachi sets.

(c) It was later discovered that 25 Rony sets sold in June 2001 had been omitted from the table of data. They had been overlooked because the company had not, after 1200 days, been called out to deal with any breakdowns of these 25 machines. Using this additional information:
(i) modify the estimates you made in part (a)
(ii) state how, if at all, your answer to part (b) would be changed.

(d) Give a reason why the median and the interquartile range were used in preference to the mean and standard deviation on times to first call-out.

Answers can be found on page 144.

Key points to remember

- If a trial can result in a number of possible **outcomes**, each outcome may have a probability associated with it.

- Probability is measured on a scale from 0 to 1. 0 represents **impossibility** and 1 represents **certainty**.

- If an outcome has probability 0.2 it is expected to occur, in the long run, in 0.2 or $\frac{1}{5}$ of all trials.

- **Events** consist of **one or more outcomes**.

- Examination questions will either give you the probability of an event, or require you to derive it using '**equally likely outcomes**'.

- The event that A does not occur is usually denoted A'. $P(A') = 1 - P(A)$.

- Two events are **mutually exclusive** if they cannot both occur as a result of the same trial.

- If A and B are mutually exclusive events **$P(A \cup B)$ = $P(A) + P(B)$**. $P(A \cup B)$ represents the probability of event A **or** event B happening. It includes the case of both happening (but this cannot occur with mutually exclusive events).

- The law may be extended to more than two mutually exclusive events.

- A and B are **independent events** if the probability of A happening is not affected by whether B happens. Independent events could be different outcomes of the same trial but are much more likely to be outcomes of different trials.

- $P(A \mid B)$ is the **conditional probability** of A happening given that B happens. A and B may occur simultaneously but it is easier to think of A happening after B.

- If A and B are independent then **$P(A \mid B) = P(A)$**. It is also true that if $P(A \mid B) = P(A)$ then A and B are independent.

- **$P(A \cap B)$ = $P(A) \times P(B \mid A)$**.

- $P(A \cap B)$ represents the probability of events A and B both occurring. If A and B are **independent**, this law simplifies to **$P(A \cap B) = P(A) \times P(B)$**.

- This law may be extended to more than two events.

Formulae you must know

- $P(A') = 1 - P(A)$

- $P(A \cup B) = P(A) + P(B)$ provided A and B are mutually exclusive.

  This may be extended to $P(A \cup B \cup C \ldots) = P(A) + P(B) + P(C) + \ldots$ provided A, B, C … are mutually exclusive.

- $P(A \cap B) = P(A) \times P(B \mid A)$

  This may be extended to $P(A \cap B \cap C \ldots) = P(A) \times P(B \mid A) \times P(C \mid A \text{ and } B) \times \ldots$

  If A, B, C … are independent this simplifies to:

  $P(A \cap B \cap C \ldots) = P(A) \times P(B) \times P(C) \ldots$

Don't make these mistakes …

Don't use a probability that is **negative or greater than 1**. If your calculations produce probabilities like this, **you have definitely made a mistake**.

When calculating the probability of **two events both occurring** don't add the probabilities. **Multiply** them – the probability of two events both occurring will be less than their individual probabilities of occurring.

When calculating the probability of **one or other of two events occurring**, don't multiply the probabilities. **Add** them – the probability of one or other event occurring will be greater than their individual probabilities of occurring.

Q1 The probability that callers to a railway timetable enquiry service receive accurate information is 0.9. Find the probability that of three randomly selected callers:

(a) all receive accurate information
(b) exactly two receive accurate information
(c) less than two receive accurate information.

(a) $0.9 \times 0.9 \times 0.9 = 0.729$

(b) $P(A) = 0.9 \quad P(I) = 1 - 0.9 = 0.1$

AAI: $0.9 \times 0.9 \times 0.1 = 0.081$
AIA: $0.9 \times 0.1 \times 0.9 = 0.081$
IAA: $0.1 \times 0.9 \times 0.9 = 0.081$

P(2 customers receiving accurate information) $= 3 \times 0.081 = 0.243$

(c) P($\geqslant$2 customers receiving accurate information)
$= 0.729 + 0.243 = 0.972$

P(<2 customers receiving accurate information)
$= 1 - 0.972 = 0.028$

- Each customer will receive either accurate or inaccurate information.
- All three must receive accurate information. Events are **independent**. **Multiply** the individual probabilities.
- Each line is the probability of a particular customer receiving inaccurate information and the other two receiving accurate information. There are three **mutually exclusive** ways in which this can happen, so you need to **add** the probabilities.
- You could calculate P(less than 2) directly but it is easier to use the results you have already found for P(not less than 2).
- Alternatively, you could use a tree diagram to show all the possible outcomes and their probabilities. You may find this method easier but it only really works for a relatively small number of alternatives.

Q2 Three persons are to be selected at random, one after another, from a group of eight persons of whom five are female and three are male. Calculate the probability that:

(a) each of the first two persons selected will be female and the third will be male
(b) two females and one male will be selected
(c) all three selected will be of the same sex.

- This question is about **conditional probability**. If the first person selected is female, that leaves seven, of whom four are female. After two females have been selected there are six left, of whom three are male.

(a) $\frac{5}{8} \times \frac{4}{7} \times \frac{3}{6} = 0.179$

(b) FFM: $\frac{5}{8} \times \frac{4}{7} \times \frac{3}{6} = \frac{5}{28}$

FMF: $\frac{5}{8} \times \frac{3}{7} \times \frac{4}{6} = \frac{5}{28}$

MFF: $\frac{3}{8} \times \frac{5}{7} \times \frac{4}{6} = \frac{5}{28}$

P(2 females and 1 male) $= 3 \times \frac{5}{28} = 0.536$

(c) FFF: $\frac{5}{8} \times \frac{4}{7} \times \frac{3}{6} = \frac{5}{28}$

MMM: $\frac{3}{8} \times \frac{2}{7} \times \frac{1}{6} = \frac{1}{56}$

P(all 3 the same sex) $= \frac{5}{28} + \frac{1}{56} = 0.196$

- Note that although the sexes are selected in different orders the probabilities of each order are the same.
- There are three mutually exclusive ways of selecting two females and one male. You need to **add** the probabilities.
- You could also have answered this question using a tree diagram.

Q3 Last year the employees of a firm received no bonus, a small bonus or a large bonus. The following table shows the number in each category, classified by whether or not they worked for the IT department.

	No bonus	Small bonus	Large bonus
Not IT dept.	25	85	5
IT dept.	4	8	23

A tax inspector decides to investigate the tax affairs of an employee selected at random.

D is the event that an employee who works for the IT department is selected.

E is the event that an employee who received no bonus is selected.

D' and E' are the events 'not D' and 'not E' respectively.

Find:

(a) P(D)   **(b)** P(D∪E)   **(c)** P(D'∩E').

F is the event that an employee is female.

(d) Given that P(F') = 0.8, find the number of female employees.

(e) Interpret P(D│F) in the context of this question.

(f) Given that P(D∩F) = 0.1, find P(D│F).

(a) $P(D) = \frac{35}{150} = 0.233$

(b) $P(D∪E) = \frac{60}{150} = 0.4$

(c) $P(D'∩E') = \frac{90}{150} = 0.6$

(d) $P(F) = 1 - 0.8 = 0.2$

There are $150 \times 0.2 = 30$ female employees.

(e) P(D│F) is the probability the employee selected works for the IT department given that she is female.

(f) $P(D∩F) = P(F) \times P(D│F)$

$0.1 = 0.2 \times P(D│F)$

$P(D│F) = 0.5$

- There are 150 employees who are each equally likely to be selected. 35 of these work for the IT department.

- 60 employees work for the IT department or received no bonus (or both).

- 90 employees do not work for the IT department and received a bonus (not 'no bonus').

Q1 Customers at a supermarket pay by cash, cheque or credit card. The probability of a randomly selected customer paying by cash is 0.64 and by cheque is 0.13.

(a) Determine the probability of a randomly-selected customer paying by credit card.

(b) Two customers are selected at random. Find the probability of:
  (i) them both paying by credit card
  (ii) one paying by credit card and the other paying by cash.

(c) Three customers are selected at random. Find the probability of:
  (i) all three paying by cash
  (ii) exactly one paying by cheque
  (iii) one paying by cash, one by cheque and one by credit card.

Q2 When Yasmin is on holiday she intends to go for a walk before breakfast each day. However, sometimes she stays in bed instead. The probability that she will go for a walk on the first morning is 0.8. Thereafter, the probability that she will go for a walk is 0.8 if she went for a walk on the previous morning and 0.4 if she did not.

Find the probability that on the first three days of the holiday she will go for:

(a) three walks

(b) exactly two walks.

Q3 The probability that an archer hits the target depends upon weather conditions.

If it is windy the probability of hitting the target is $\frac{2}{5}$ and if it is not windy the probability is $\frac{4}{5}$. The probability that it is windy on a random day in June is $\frac{1}{6}$.

(a) Calculate the probability that the archer hits the target on a random day in June.

(b) Given that the archer hits the target on a day in June, find the probability that it was a windy day.

Q4 A refreshment stall at a summer fair sells spring water, orange juice and lemonade. The probability that a randomly selected customer will choose spring water is 0.2, orange juice is 0.35 and lemonade is 0.45.

Find the probability that three randomly selected customers all choose:

(a) spring water

(b) the same drink

(c) spring water, given that they all choose the same drink.

Q5 A firm recruits 120 sales staff on a three-month trial. At the end of three months, they are either offered permanent employment, offered a further trial or their employment is terminated.

	Age (years)		
	< 20	20–25	> 25
Permanent employment	47	20	13
Further trial	17	3	1
Employment terminated	11	7	1

One of the recruits is selected at random.

Q denotes the event that the age of the selected recruit is < 20 years.
R denotes the event that the selected recruit was offered permanent employment.
S denotes the event that the selected recruit was offered a further trial.
(Q', R' and S' denote the events not Q, not R and not S respectively.)

Determine the value of:

(a) $P(R)$

(b) $P(Q \cap R)$

(c) $P(Q \cup S')$

(d) $P(R \mid Q')$.

It is known that 45 out of the 120 recruits are female.

F denotes the event that the selected recruit is female.

(e) Find $P(F \cap R)$ given that F and R are independent.

(f) Find $P(F \cap Q)$ given that $P(Q \mid F) = 0.8$.

(g) Find $P(S \mid F)$ given that $P(S \cap F) = 0.05$.

Q6 It is known that 1% of the population suffers from a certain disease. A diagnostic test for the disease gives a positive response with probability 0.98 if the disease is present. If the disease is not present, the probability of a positive response is 0.005.

(a) A test is applied to a randomly selected person.

(i) Show that the probability of this test giving a positive response is 0.014 75.

(ii) Given that the test gave a positive response, calculate the probability that the person has the disease.

(b) A randomly-selected person is tested and a positive response is obtained. This person is tested again. Assuming that the tests are independent, calculate the probability that this second test will give a positive response.

Answers can be found on page 145.

19 Binomial distribution

- The **binomial distribution** is used for trials that have **two possible outcomes**. For example a commuter train either arrives on time or is late.

- The outcomes may be called '**success**' and '**failure**'.

- If the probability of 'success' in each trial is p then the probability of exactly r 'successes' in n independent trials is:

  $$\binom{n}{r}p^r(1-p)^{n-r} \quad \text{where} \binom{n}{r} = \frac{n!}{r!(n-r)!}$$

  For the binomial distribution to apply the trials must be **independent** and p must be constant.

- Either outcome may be defined as 'success' but the usual convention is to call the outcome with the smaller probability 'success'. Thus p will be $\leqslant 0.5$.

- The distribution is **discrete** and the possible outcomes are 0, 1, 2, ... , n.

- The binomial distribution has **mean** np and **variance** $np(1-p)$.

- The **standard deviation** is $\sqrt{np(1-p)}$ and is a more useful measure of spread than the variance.

- Tables of the binomial distribution list the probabilities of r or fewer successes in n trials. Although you should use these tables where possible, not all values of n and p can be included.

- Some calculators will calculate binomial probabilities directly.

- The notation **B(10, 0.2)** denotes a binomial distribution with $n = 10$ and $p = 0.2$.

Don't make these mistakes ...

Don't assume that any question involving trials with two outcomes may be answered using the binomial distribution. The distribution also assumes:

- a fixed number of trials
- the trials are independent
- p is constant.

Formulae you must know

- $P(R = r) = \binom{n}{r}p^r(1-p)^{n-r}$

- $0! = 1$ and therefore $\binom{n}{n} = 1$ and $\binom{n}{0} = 1$

Q1 A pottery produces large quantities of drinking mugs and reckons that the probability of each mug being classed as 'seconds' is 0.12.

A random sample of six mugs is taken from the production. Find the probability that this sample will contain:

(a) exactly two 'seconds'
(b) two or fewer 'seconds'
(c) more than one 'second'.

(a) As this is a binomial distribution, with
$n = 6$, $p = 0.12$:

$P(R = 2)$

$= \dfrac{6.5.4.3.2.1}{2.1.4.3.2.1} \times (0.12)^2 \times (1 - 0.12)^{6-2}$

$= 0.129\,53 = 0.130$

(b) $P(R \leq 2) = P(0) + P(1) + P(2)$

$= 0.88^6 + 6 \times 0.12 \times 0.88^5 + 0.129\,53$

$= 0.974$

(c) $P(R > 1) = 1 - P(0) - P(1)$

$= 1 - 0.88^6 - 6 \times 0.12 \times 0.88^5$

$= 0.156$

- There are two possible outcomes of each trial. The question implies that p is **constant** and the trials are **independent**.

- Your tables are very unlikely to include B(6, 0.12) so you will have to use the formula to evaluate the probabilities.

- It is sensible to give the answer to **three significant figures**, but to keep five for use in calculations later in the question.

- You have already calculated P(2) so you don't need to calculate it again.

- You can use three significant figures here, as you will not be using this result later.

- This is much quicker than calculating P(2) + P(3) + P(4) + P(5) + P(6).

Q2 The probability that any A-level candidate will be absent at the start of an examination is 0.01, independently of whether other candidates are absent. Find the probability that, of 50 A-level candidates, the number absent at the start of the examination will be:

(a) 3 or fewer   (b) exactly 3
(c) 3 or more   (d) more than 1.

(a) As this is a binomial distribution,
with $n = 50$, $p = 0.01$:

$P(3 \text{ or fewer}) = 0.9984$

(b) $P(3) = P(3 \text{ or fewer}) - P(2 \text{ or fewer})$

$= 0.9984 - 0.9862 = 0.0122$

(c) $P(3 \text{ or more}) = 1 - P(2 \text{ or fewer})$

$= 1 - 0.9862 = 0.0138$

(d) $P(\text{more than } 1) = 1 - P(1 \text{ or fewer})$

$= 1 - 0.9106 = 0.0894$

- The question states that the two possible outcomes of each trial are independent.

- It is possible, but very time consuming to calculate the answers as in the previous question. You will be provided with tables of B(50, 0.01) so save time by using them.

- P(r or fewer) can be **read directly from the tables**.

- You could round to three significant figures but as the answer is so close to 1 it is sensible to give four significant figures.

- In order to use the tables you need to express the required probability in terms of 'P(r or fewer)'. Be particularly careful about the difference between 'r or more' and 'more than r'.

Q3 A motorist passes through a set of traffic lights on the way to work each weekday morning. The probability that the motorist passes through the traffic lights without needing to stop is 0.3.

(a) Find the probability that, on 20 consecutive weekday mornings the motorist passes through the traffic lights without needing to stop:
   (i) 4 or more times
   (ii) exactly five times.

The motorist also passes through the traffic lights on Saturday and Sunday on the way to visit an elderly relative.

(b) State, giving a reason, whether or not it is likely that the following random variables may be modelled by a binomial distribution.
   (i) *Y*, the number of times the motorist passes through the traffic lights without needing to stop on 20 consecutive days (i.e. including Saturday and Sunday)
   (ii) *Z*, the number of consecutive weekdays before the motorist passes through the traffic lights without needing to stop on four occasions.

(a) (i) As this is a binomial distribution with $n = 20$, $p = 0.3$:

P(4 or more) = 1 − P(3 or fewer)

= 1 − 0.1071

= 0.893

(ii) P(5) = P(5 or fewer) − P(4 or fewer)

= 0.4164 − 0.2375

= 0.179

(b) (i) This is not a binomial distribution as *p* is not constant.
   (ii) This is not a binomial distribution, as *n* is not constant.

- There are two possible outcomes of each trial.
- The probabilities of needing to stop on different days are independent.
- You will have tables of B(20, 0.3), so you can save time by using them.

- There are two possible outcomes but the probability of having to stop will be different at weekends than on weekdays.
- The trials continue until the motorist passes through the traffic lights without needing to stop 4 times.
   For a **binomial** the **number of trials must be known** in advance.

Q1 Leaflets advertising a pop concert are handed out to students in a large city. For each student who receives a leaflet, the probability of subsequently attending the concert is 0.09.

Find the probability that, from a sample of 20 students who received a leaflet, the number who subsequently attend the concert will be:

(a) 2 or fewer

(b) exactly 4.

Q2 Vehicles approaching a T-junction must either turn right or turn left. It is observed that 43% turn right. Find the probability that of a random sample of eight vehicles, approaching the junction, the number turning right will be:

(a) exactly four

(b) fewer than two

(c) six or fewer.

Q3 A golfer practises on a driving range. He counts 'success' as driving a ball within 15 m of the flag. The probability of 'success' with each particular drive is 0.3.

If he drives ten balls, find the probability of

(a) four or fewer 'successes'

(b) exactly four 'successes'

(c) from two to five (inclusive) 'successes'

(d) four or fewer 'failures'.

Find the mean and standard deviation of the number of 'successes' in 20 drives.

Q4 Items from a production line are examined for any defects. The probability that any item will be found to be defective is 0.15, independently of all other items.

(a) A batch of 16 items is inspected. Using tables of cumulative binomial probabilities, or otherwise, find the probability that:
(i) at least four items in a batch are defective
(ii) exactly four items in a batch are defective.

(b) Five batches, each containing 16 items, are taken.
(i) Find the probability that at most two of these five batches contain at least four defective items.
(ii) Find the expected number of batches that contain at least four defective items.

Q5 In the first round of a longjump competition each competitor makes four jumps. If the competitor oversteps a line the jump is disallowed. The numbers of jumps disallowed for the 40 competitors are summarised below.

Number of jumps disallowed	0	1	2	3	4
Number of competitors	14	9	5	6	6

(a) Calculate the mean and standard deviation of the number of jumps disallowed per competitor.

(b) Calculate the proportion p of jumps disallowed.

(c) Assuming the number of jumps disallowed for each competitor follows a binomial distribution, use your calculated value of p to estimate the mean and variance of this distribution.

(d) Do you think the binomial distribution is an adequate model for the data above? Give a reason.

(e) Give a reason, apart from any numerical calculations, why the number of jumps disallowed may not follow a binomial distribution.

Q6 A doctor wishes to undertake a trial into the effectiveness of a new treatment for a skin condition. She asks patients to take part in the trial and observes that the probability of them agreeing is 0.3 and may be assumed to be independent of whether or not other patients agree.

(a) If she asks 25 patients to take part in the trial, find the probability that:
   (i) four or fewer will agree
   (ii) exactly 19 will **not** agree.

The probability that a patient who agrees to take part in the trial withdraws before the end is 0.23.

(b) For **each** of the following cases state giving a reason, whether the binomial distribution is likely to provide an adequate model for the random variable R.
   (i) Out of 100 patients taking part in the trial, R is the total number withdrawing before the end.
   (ii) R is the total number of patients asked, in order to obtain 100 to take part in the trial.

Answers can be found on page 145.

Key points to remember

- The normal distribution is a **continuous** distribution.
- Probability is represented by the **area under a curve**.

  (i) The total area under this curve is 1.
  (ii) The curve is bell-shaped.
  (iii) The curve is symmetrical.

- Normal distribution tables list a **standard normal distribution**, that is a normal distribution with mean 0 and standard deviation 1. This is usually denoted z.

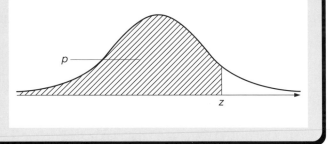

$P(a < x < b)$

- To make calculations you must:

  - **either** standardise x, an observation from a normal distribution with mean μ and standard deviation σ using:

  $$z = \frac{x - \mu}{\sigma}$$

  - **or** convert standard normal values to x by rearranging the formula:

  $$x = \mu + z\sigma$$

- Tables enable you to find p for a given value of z, and z for a given value of p.

- **Remember** that the standard normal distribution has mean zero. Negative values of z are in the lower half of the distribution and positive values of z are in the upper half.

Formulae you must know

- $z = \dfrac{x - \mu}{\sigma}$

Don't make these mistakes ...

Never subtract x from the population mean μ, **always subtract μ from x**.

Don't try to work out in your head how to combine probabilities. Always draw a diagram.

Q1 Tins of peas are filled and sealed by a machine.

The weight of peas in a tin is normally distributed with mean 435 g and standard deviation 25 g. Find the probability that the weight of the peas in a randomly selected tin:

(a) is less than 470 g
(b) exceeds 465 g
(c) lies between 410 g and 430 g.

- Note $\mu = 435$, $\sigma = 25$

(a) $z = \dfrac{470 - 435}{25} = 1.4$

$P(X < 470) = 0.919$

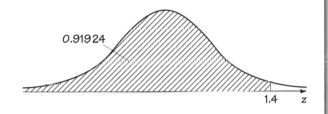

- **Standardise** 470 using $z = \dfrac{(x - \mu)}{\sigma}$.

- Draw a diagram and **identify the area** you need.

- The shaded area is given by your tables.

- The tables give five places of decimals, but it is sensible to round your final answer to three significant figures.

(b) $z = \dfrac{465 - 435}{25} = 1.2$

$P(X > 455) = 1 - 0.884\,93 = 0.115$

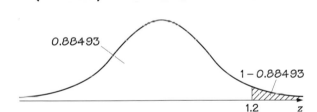

- **Standardise** 465.

- You can use the fact that the **total area under the curve is 1** to find the required area.

(c) $z_1 = \dfrac{410 - 435}{25} = -1.0$

$z_2 = \dfrac{430 - 435}{25} = -0.2$

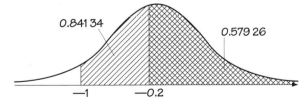

- **Standardise** 410 and 430. Note that the **negative signs** mean that the z-values are in the **lower half of the distribution**.

- Now use **symmetry**. The area above -1 is the same as below 1.

- From the diagram, you should see that the area you need is the difference between the two areas you found from tables.

- Keep as many significant figures as possible in the calculation but round the final answer to three significant figures.

$P(410 < X < 430)$
$= 0.841\,34 - 0.579\,26 = 0.262$

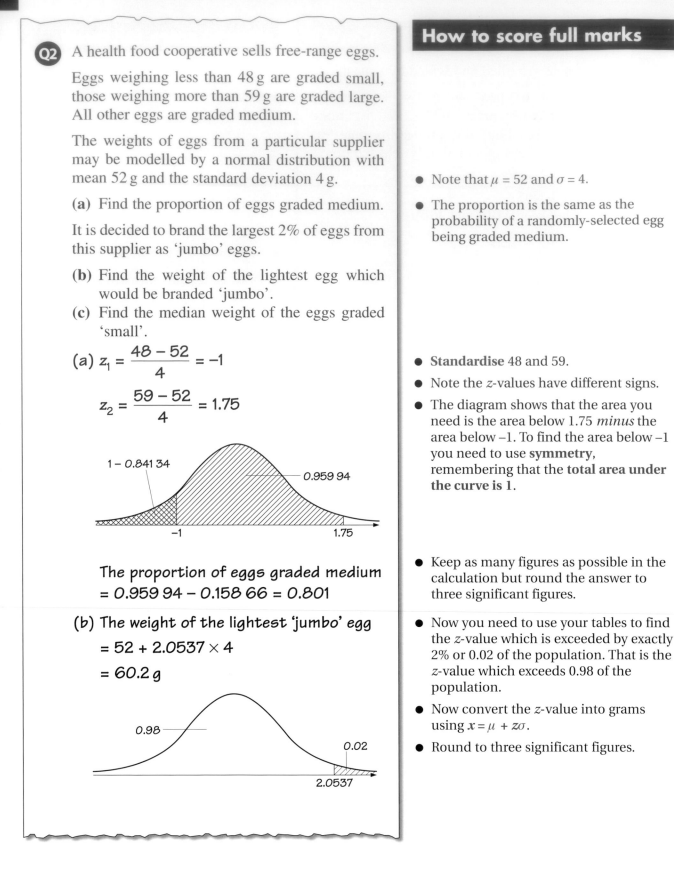

Q2 A health food cooperative sells free-range eggs.

Eggs weighing less than 48 g are graded small, those weighing more than 59 g are graded large. All other eggs are graded medium.

The weights of eggs from a particular supplier may be modelled by a normal distribution with mean 52 g and the standard deviation 4 g.

(a) Find the proportion of eggs graded medium.

It is decided to brand the largest 2% of eggs from this supplier as 'jumbo' eggs.

(b) Find the weight of the lightest egg which would be branded 'jumbo'.

(c) Find the median weight of the eggs graded 'small'.

(a) $z_1 = \dfrac{48 - 52}{4} = -1$

$z_2 = \dfrac{59 - 52}{4} = 1.75$

The proportion of eggs graded medium
= 0.959 94 − 0.158 66 = 0.801

(b) The weight of the lightest 'jumbo' egg
= 52 + 2.0537 × 4
= 60.2 g

- Note that $\mu = 52$ and $\sigma = 4$.

- The proportion is the same as the probability of a randomly-selected egg being graded medium.

- **Standardise** 48 and 59.

- Note the z-values have different signs.

- The diagram shows that the area you need is the area below 1.75 *minus* the area below −1. To find the area below −1 you need to use **symmetry**, remembering that the **total area under the curve is 1**.

- Keep as many figures as possible in the calculation but round the answer to three significant figures.

- Now you need to use your tables to find the z-value which is exceeded by exactly 2% or 0.02 of the population. That is the z-value which exceeds 0.98 of the population.

- Now convert the z-value into grams using $x = \mu + z\sigma$.

- Round to three significant figures.

(c) From part (a), the smallest 0.158 66 of the eggs are graded small. The median of these eggs will exceed half of 0.158 66 or 0.079 33 of the population.

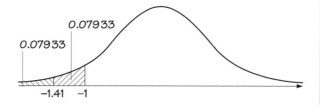

Median of eggs graded small is
52 − 1.41 × 4 = 46.4 g

Q3 A group of students believes that the time taken to travel to college, T minutes, can be assumed to be normally distributed. Within the college 5% of students take at least 55 minutes to travel to college and 0.1% take less than 10 minutes.

Find the mean and standard deviation of T.

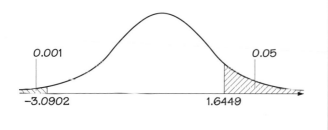

$\mu + 1.6449\sigma = 55$

$\mu - 3.0902\sigma = 10$

Solving these equations gives $\mu = 39.4$ minutes, $\sigma = 9.50$ minutes.

How to score full marks

- Use your tables to find the z-value exceeded by 0.920 67 of the population.
- Note this must be negative.
- You could increase accuracy by interpolation but it is much easier and sufficiently accurate to round the proportion to 0.92.
- Now convert the z-value into grams.

- Find the z-value exceeded by 99.9% or 0.999 of the population, and the z-value which exceeds 95% or 0.95 of the population.
- Take care with the signs.
- **Use $x = \mu + z\sigma$** (x, z known; μ, σ unknown) to obtain two **simultaneous equations**. Then solve them.

Questions to try

Q1 The random variable X has a normal distribution with a mean of 36 and a standard deviation of 2.5. Find $P(X < 32)$.

Q2 The lengths of components produced by a machine are normally distributed with a mean of 1.984 cm and a standard deviation of 0.006 cm. To function satisfactorily a component should measure between 1.975 cm and 1.996 cm in length. Find the probability that a randomly-selected component will function satisfactorily.

Q3 A test carried out on the blood of patients in intensive care results in a measurement which, for smokers, may be modelled by a normal random variable with mean 310 and standard deviation 110.

(a) What proportion of smokers have a measurement lower than 250?
(b) What measurement is exceeded by 20% of smokers?

Q4 The lifetime of a particular safety device is found to be normally distributed with mean 15200 hours and standard deviation 400 hours.

Calculate, to three significant figures:

(a) the probability that a safety device selected at random has a lifetime less than 15800 hours
(b) the proportion of such safety devices having lifetimes between 14500 and 15800 hours
(c) the 84th percentile of this distribution.

Q5 A swimming pool manager reported that the time spent in the pool by a user could be modelled by a normal distribution with mean 70 minutes and standard deviation 20 minutes.

(a) Assuming that this model is adequate, what is the probability that a user spends:
  (i) less than 95 minutes in the pool
  (ii) between 65 and 95 minutes in the pool?

The pool closes at 9.00 p.m.

(b) Explain why the model above could not apply to a user who entered the pool at 8.00 p.m.
(c) Estimate an approximate latest time of entry for which the model above could still be plausible.

Q6 A clothes manufacturer estimated that adult male customers requiring jackets will have chest measurements which may be modelled by a normal distribution with mean 102 cm and standard deviation 5 cm. It is decided to make three sizes:

large to fit chests over 109 cm
medium to fit chests between 99 cm and 109 cm
small to fit chests of less than 99 cm.

(a) Assuming the model is adequate, calculate the proportion of adult male customers who will require:
  (i) **large** jackets
  (ii) **medium** jackets.

It is decided to introduce an **extra large** size to fit the largest 2% of chests of adult male customers.

(b) Find the minimum chest measurement for **extra large** jackets.
(c) Find the median chest measurement of those customers who require **small** jackets.

Answers can be found on page 146.

21 Correlation and regression

Key points to remember

- **Correlation** and **regression** are about **relationships** between two variables.

- **Simple linear regression** uses observed data to estimate an equation of the form $y = a + bx$ which relates a response (or dependent) variable, y, to an explanatory variable x. This is known as the **regression of y on x**.

- The variables are not interchangeable. For example, the depth of a river, y, may depend on the previous week's rainfall, x. The previous week's rainfall cannot depend on the depth of the river. Although it is possible to calculate the regression equation of rainfall on depth of river, it is meaningless. You should only calculate the regression equation of depth of river on rainfall.

- The equation may be used to estimate the value of y corresponding to a particular value of x, but only if the relationship is approximately linear and the value of x is within the range of the observed values.

- The **product moment correlation coefficient** measures the strength of a linear relationship between two variables. It is measured on a scale from -1 to $+1$.

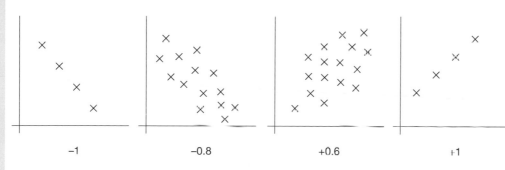

- In correlation the variables x and y are interchangeable.

- **Spearman's rank correlation** coefficient may be calculated by replacing the raw values of x and y by their ranks and then calculating the product moment correlation coefficient between ranks.

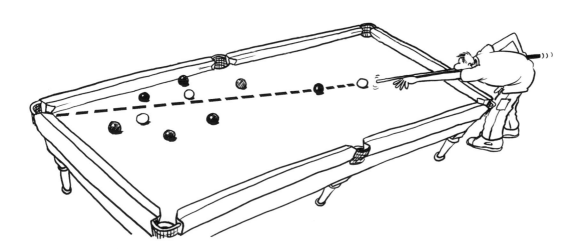

Formulae you must know

Provided you have a calculator with statistical functions there are no formulae you must know. The regression and correlation coefficients may be obtained directly from your calculator – at least in real life. If your examination board does not give you the raw data, but only summations, you may need some of these formulae.

Given n pairs of observed values (x, y):

for the regression equation $y = a + bx$

$$\bullet\ b = \frac{\Sigma(x - \bar{x})(y - \bar{y})}{\Sigma(x - \bar{x})^2}$$

which may be calculated directly or by using the formula:

$$\bullet\ b = \frac{\Sigma xy - \frac{\Sigma x \Sigma y}{n}}{\Sigma x^2 - \frac{(\Sigma x)^2}{n}}$$

$$\bullet\ a = \bar{y} - b\bar{x}$$

The product moment correlation coefficient, r, is given by:

$$\bullet\ r = \frac{\Sigma(x - \bar{x})(y - \bar{y})}{\sqrt{\Sigma(x - \bar{x})^2 \Sigma(y - \bar{y})^2}}$$

which may be calculated directly or by using the formula:

$$\bullet\ r = \frac{\Sigma xy - \frac{\Sigma x \Sigma y}{n}}{\sqrt{\left(\Sigma x^2 - \frac{(\Sigma x)^2}{n}\right)\left(\Sigma y^2 - \frac{(\Sigma y)^2}{n}\right)}}$$

Spearman's rank correlation coefficient may be found by using ranks in the formula for r. An alternative formulation is:

$$\bullet\ r_s = 1 - \frac{6\Sigma d^2}{n(n^2 - 1)}$$

where d is the difference between the x-rank and the y-rank for an observation. This gives the same result as the product moment correlation coefficient between ranks, provided there are no tied ranks. If there are tied ranks the result will be slightly different. In this case the correct formula to use is the product moment correlation coefficient between ranks. However you are unlikely to be penalised in an examination for using a different formula.

Don't make these mistakes...

Don't calculate the regression equation of p on q if the variable q depends on the variable p. Instead, calculate the regression equation of q on p.

Don't extrapolate – that is, don't use the regression equation of y on x to predict y for values of x outside the range of the observed values of x.

Don't calculate the product moment correlation coefficient for variables which are not approximately linearly related or which contain outliers. It may be appropriate to calculate Spearman's rank correlation coefficient in these circumstances.

Don't claim that changes in x cause changes in y if your only evidence is a high correlation coefficient between x and y. This shows that the variables are associated (perhaps through another variable) but not that changing one causes the other to change.

Q1 The following data were extracted from a daily newspaper for eight days in October.

Rainfall, x cm	1.3	3.8	4.2	2.6	2.1	2.6	5.3	0.9
Sunshine, y hours	1.5	0.3	0.0	4.2	3.6	0.5	0.0	1.4

(a) Calculate Spearman's rank correlation coefficient for the data.

(b) Comment on the value you have obtained.

(a)

Rank, x	7	3	2	4.5	6	4.5	1	8
Rank, y	3	6	7.5	1	2	5	7.5	4

From the calculator, $r_s = -0.711$

- The data have been ranked from largest to smallest. They could have been ranked from smallest to largest, but, in order to interpret the coefficient, both x and y should be ranked in the same direction.

- One rainfall of 2.6 cm should be ranked 4 and the other ranked 5. From the given data it is not possible to decide which is larger and so both have been given the mean of 4 and 5, i.e. 4.5.

- This is the **product moment correlation coefficient** between the ranks.

- The formula based on Σd^2 gives −0.690 in this case. The relatively large difference is due to the relatively large proportion (25%) of tied ranks.

(b) The correlation coefficient shows some tendency for days with the most rainfall to have the least sunshine.

- The **negative sign** shows high rainfall associated with low sunshine. The **magnitude** shows that the association is moderately strong.

Q2 **(a)** Estimate, without undertaking any calculations, the value of the product moment correlation coefficient in each of the scatter diagrams below.

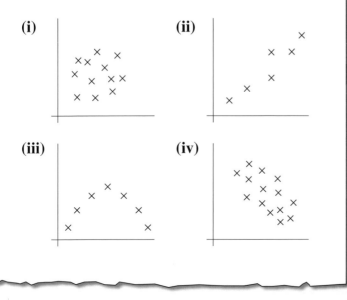

(i)

(ii)

(iii)

(iv)

(b) Why is the product moment correlation coefficient an unsuitable measure to use for diagram (iii)?

(a) (i) 0.0
    (ii) 0.95
    (iii) 0.0
    (iv) −0.5

(b) A non-linear relationship is indicated.

How to score full marks

- If the data show an **upward trend** the coefficient will be **positive**, if they show a **downward trend** the coefficient will be **negative**.

- The closer to a straight line the points lie, the closer the magnitude of the coefficient is to 1.

- If you imagine moving the origin to $(\bar{x}, \bar{y})$ then points that lie in the first and third quadrants will make a positive contribution and points that lie in the second and fourth quadrants will make a negative contribution.

Q3 **(a)** Suppose you were asked to relate the variables r and s with a regression line of the form $y = a + bx$ where r is the population of a country and s is the number of medals won at the Sydney Olympic Games.
> **(i)** State, giving a reason, which of the variables you would choose for x and which for y.
> **(ii)** Would you expect b to be positive or negative?

(b) Repeat part (a) if:
> r is the temperature of a cup of tea and
> s is the time which has elapsed since it was poured out.

(a) (i) $r \rightarrow x$ and $s \rightarrow y$ because the number of medals won may depend on the size of the population, but the size of the population cannot depend on the number of medals won.

    (ii) Positive – it is likely that countries with larger populations will win more medals.

(b) (i) $s \rightarrow x$ and $r \rightarrow y$ because the temperature will depend on how long the cup of tea has been poured out, the time cannot depend on the temperature of the cup of tea.

    (ii) Negative, as time passes the temperature will reduce.

Q4 The table below shows x, a measure of social deprivation of an area (the higher the value of x, the more deprived the area) and y, the percentage of pupils at the local comprehensive school gaining four or more GCSE passes at grade C or above.

School	A	B	C	D	E	F	G	H	I	J	K
x	3.5	28.3	17.9	4.6	12.2	13.5	53.2	67.1	36.2	14.7	26.4
y	78.7	51.3	57.5	72.9	70.4	52.4	41.2	19.0	44.9	64.4	48.8

(a) Draw a scatter diagram of the data.

(b) Calculate the regression line of y on x and draw it on your diagram.

(c) The headteacher of school F claims that it is a better school than G. Comment on this claim.

(a)

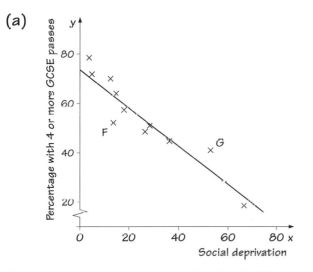

(b) Regression line is $y = 74.7 - 0.791x$
$x = 0, y = 74.7$ $x = 60, y = 27.2$

(c) The diagram clearly shows an association between social deprivation and percentage GCSE passes. School F has a higher percentage pass rate than School G. However, School F is below the regression line indicating a lower pass rate than would be expected from its social deprivation index. School G is above the regression line indicating a higher percentage pass rate than would be expected from its social deprivation index. Even if the percentage of GCSE passes is accepted as a valid measure of a school's worth, there is no basis for the headteacher's claim.

- You can find these values from the calculator. It is reasonable to use three significant figures. To draw the line, calculate the coordinates of two points at opposite ends of the graph and join them. You should calculate a third point as a check.

- If the line does not pass through the points plotted on the scatter diagram, you have made a mistake. Check your calculation of the regression line.

- Look for comments which relate to the data you have been given.

- It is OK to make general comments, such as questioning the value of the measures used, in addition to but not instead of making comments based on the data.

Questions to try

Q1 During the lambing season, eight ewes and the lambs they bore were weighed at the time of birth.

Ewe	A	B	C	D	E	F	G	H
Weight of ewe, kg	49	46	48	45	46	42	43	40
Weight of lamb, kg	3.7	3.0	3.4	2.9	3.1	2.7	3.0	2.8

Calculate the product moment correlation coefficient between the weights of the ewes and the weights of their lambs.

Comment, briefly, on your result.

Q2 As part of a practical exercise in statistics, Miriam was shown photographs of 11 people and asked to estimate their ages. The actual ages and the estimates made by Miriam are shown below.

Actual age, x	86	55	28	69	45	7	17	11	37	2	78
Miriam's estimate, y	88	60	35	77	50	8	15	6	49	2	85

(a) Draw a scatter diagram of Miriam's estimate, y, and the actual age, x.

(b) Calculate the equation of the regression line of Miriam's estimate on actual age.

(c) Draw this regression line on your scatter diagram. Draw also the line $y = x$.

Comment on Miriam's estimates.

Q3 The following data show the GDP per capita, x, (in US$) and the mortality before age 5, y, (per thousand live births) for a sample of 11 countries.

Country	A	B	C	D	E	F	G	H	I	J	K
x	390	17850	1680	6030	5610	510	1170	1740	2460	19860	11400
y	150	43	121	53	41	169	143	59	75	20	39

(a) Draw a scatter diagram of the data. Describe the relationship between GDP per capita and mortality before age 5 suggested by the diagram.

(b) An economist asks you to calculate the product moment correlation coefficient.
    (i) Carry out this calculation.
    (ii) Explain briefly to the economist why this calculation may not be appropriate.

(c) Calculate Spearman's rank correlation coefficient for the data.

(d) Compare and comment on the results of your calculations in parts (b)(i) and (c).

Q4 A drilling machine can run at various speeds, but in general the higher the speed the sooner the drill needs to be replaced. Over several months, 15 pairs of observations relating to speed, s revolutions per minute, and life of drill, h hours, are collected.

For convenience the data are coded so that $x = s - 20$ and $y = h - 100$ and the following summations obtained.

$\Sigma x = 143$; $\Sigma y = 391$; $\Sigma x^2 = 2413$; $\Sigma y^2 = 22\,441$; $\Sigma xy = 484$

(a) Find the equation of the regression line of h on s.

(b) Interpret the slope of your regression line.

(c) Estimate the life of a drill revolving at 30 revolutions per minute.

Q5 Ann (A), Benji (B), Chandra (C) and Dan (D) are four drivers who work for a large supermarket group. Their job is to deliver goods from a central depot to individual stores. The following table shows the load carried and the diesel consumption for ten journeys. The driver is also identified.

Driver	A	B	D	A	D	C	C	A	D	B
Load x (kg)	5650	10 100	7800	8450	5500	6950	7600	8300	6250	6600
Diesel consumption y (km/litre)	7.22	6.18	6.25	6.49	7.01	6.99	6.89	6.42	6.77	7.11

(a) Draw a scatter diagram of the data.

(b) Calculate the equation of the regression line of y on x and draw it on your scatter diagram.

(c) Interpret, in context, the slope and the intercept of the regression line.

(d) Why would it be unwise to use the regression equation to predict the diesel consumption if the load was 30 000 kg?

(e) Comment on the diesel consumption of lorries driven by Dan.

(f) Why was the regression line of y on x calculated rather than the regression line of x on y?

Answers can be found on pages 146–147.

Key points to remember

- An **algorithm** is a systematic process of well-defined, finite steps of instructions used to solve a problem.

- A **heuristic algorithm** produces a good solution to a problem using logical and intuitive steps.

- The **efficiency of an algorithm** is measured in terms of its speed of operation and the required data storage needed. To compare the efficiency of sorting algorithms you need to consider (i) the number of comparisons and (ii) the number of interchanges.

Don't make these mistakes ...

Don't mix up the shuttle sort and the bubble sort!

Don't forget in first-fit bin filling to offer each piece of data first to bin 1, then to bin 2 and so on until it fits.

Algorithms you must know

Algorithms for sorting:

- **Bubble sorts** work by placing one number in its correct position in the list in each pass.

- **Quicksort** methods take one of the numbers as a pivot and then separate the numbers into two subsets – those larger than the pivot and those smaller than the pivot.

- In an **interchange sort**, the largest number (or smallest number for ascending order sorts) is interchanged with the first number in the list.

- **Shuttle sort** methods work by comparing pairs of numbers and exchanging them if necessary.

Algorithms for searching:

- A **linear search** is an exhaustive method that checks each piece of data in turn.

- For a **sequential search**, the data is first ordered and subdivided into lists and an extra list called the index. First search the index for the correct subdivision then carry out a linear search in that subdivision to find the item.

- In a **binary search**, the data is ordered and the middle item is chosen. If this is not the required item, then the half of the list which contains the item is selected and the middle item is chosen. This method of subdividing into half lists is continued until the middle item is the required one.

Q1 Use a shell sort algorithm to rearrange the following numbers into ascending order, showing the new arrangement after each pass.

14, 27, 23, 36, 18, 25, 16, 66

14	27	23	36	18	25	16	66
a	b	c	d	a	b	c	d

14	25	16	36	18	27	23	66
a	b	a	b	a	b	a	b

14	25	16	27	18	36	23	66

14	16	18	23	25	27	36	66

- Show the **subsets** by using symbols or letters. Here we begin with INT($\frac{8}{2}$) = 4 subsets. Sort each subset using the **shuttle sort algorithm**.

- Now we use INT($\frac{4}{2}$) = 2 subsets and again shuttle sort each subset.

- Finally use shuttle sort on the full list.

Q2 25, 22, 30, 18, 29, 21, 27, 21

The list of numbers above is to be sorted into descending order.

(a) (i) Perform the first pass of a bubble sort, giving the state of the list after each exchange.

(ii) Perform further passes, giving the state of the list after each pass, until the algorithm terminates.

The numbers represent the lengths, in cm, of pieces to be cut from rods of length 50 cm.

(b) (i) Show the result of applying the first-fit decreasing bin packing algorithm to this situation.

(ii) Determine whether your solution to (b)(i) has used the minimum number of 50 cm rods.

(a)
```
25  22  30  18  29  21  27  21
25  30  22  18  29  21  27  21
25  30  22  29  18  21  27  21
25  30  22  29  21  18  27  21
25  30  22  29  21  27  18  21
25  30  22  29  21  27  21  18
30  25  29  22  27  21  21  18
30  29  25  27  22  21  21  18
30  29  27  25  22  21  21  18
```

- Show the **pairs of numbers to be exchanged**, this ensures the Method marks.

- Now you just need to show the outcome of each pass.

(b) (i) rod 1   30   18
         rod 2   29   21
         rod 3   27   22
         rod 4   25   21

(ii) total length of pieces is 193 cm
      need a minimum of 193/50 = 3.86
      i.e. 4 rods

Q3

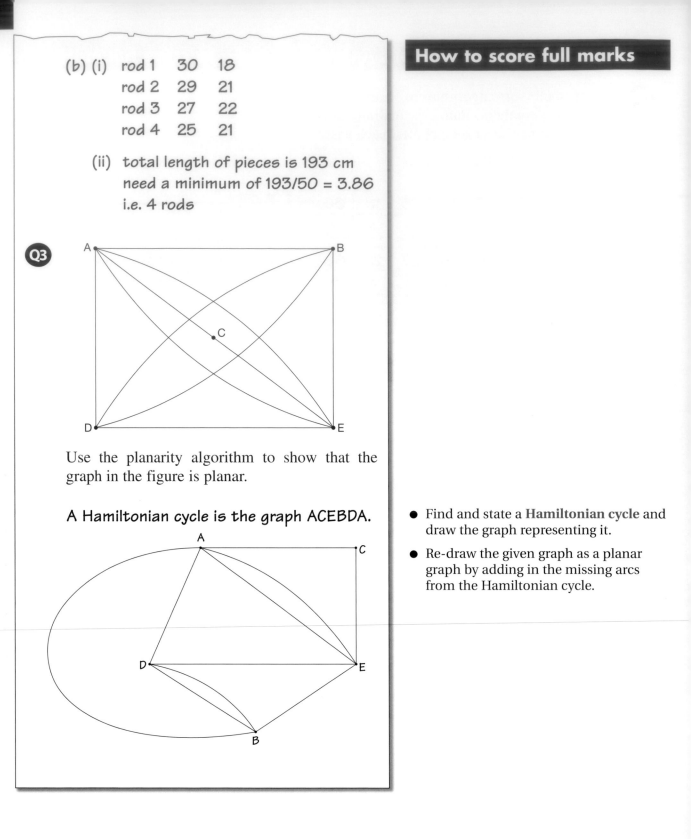

Use the planarity algorithm to show that the graph in the figure is planar.

A Hamiltonian cycle is the graph ACEBDA.

- Find and state a **Hamiltonian cycle** and draw the graph representing it.

- Re-draw the given graph as a planar graph by adding in the missing arcs from the Hamiltonian cycle.

Q1 Use the quicksort algorithm to rearrange the following list of numbers into ascending order:

63, 32, 70, 26, 59, 41, 17

Indicate entries that you have used as pivots.

Q2

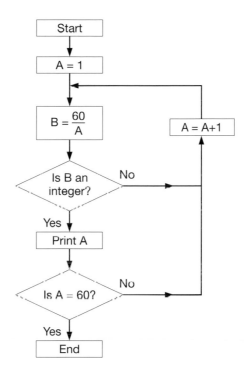

Implement the algorithm given by the flowchart above and state what the algorithm actually produces.

Q3

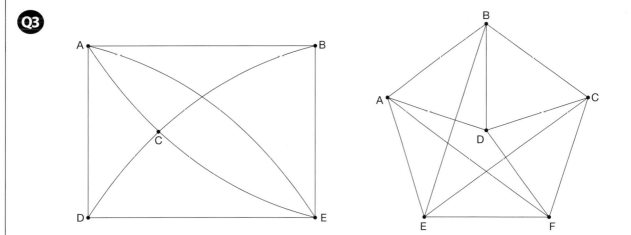

Identify if the graphs in the figures above are planar, redrawing them where necessary. Check your results using Euler's formula.

Q4 **(i)** Use the binary search algorithm to try to locate the name SUSIE in the following alphabetical list. Explain each step of the algorithm.

1.   ALLAN
2.   BRIAN
3.   DAVID
4.   EUN-JUNG
5.   FRED
6.   JOYCE
7.   HO-YI
8.   REBECCA
9.   SARAH
10. SUSIE
11. WILLIAM

(ii) Find the maximum number of iterations of the binary search algorithm needed to locate a name in a list of 10000 names.

Q4 90, 35, 105, 90, 120, 45, 60, 100

(a) Sort the list of numbers above into descending order using the bubble sort algorithm.

The numbers represent the lengths of television programmes, in minutes, to be recorded onto 180-minute video tapes.

(b) Apply the full-bin algorithm and the first-fit decreasing bin packing algorithm to this situation. Compare the results.

Answers can be found on pages 148–149.

Key points to remember

- A **graph** is a diagram showing how objects are related to each other. A **vertex** or **node** is a point representing an object in a graph; an **edge** or **arc** is a line joining two vertices (or nodes). A **path** is a route through a graph.

- A **cycle** is a path that completes a loop and returns to its starting point. An **Eulerian cycle** includes every edge (or arc) of the graph just once. A **Hamiltonian cycle** passes through every vertex (or node) once, and only once, and returns to its starting point.

- A **spanning tree** connects all the vertices in a graph or network with no cycles.

- A **network** is a graph in which every edge has a value called its **weight**.

- In a network flow problem the edges in the network have given directions representing the flow along the edge.

- The **travelling salesman problem** (TSP) attempts to find a Hamiltonian cycle of minimum length. There is no simple algorithm for solving the TSP. The strategy to solve the TSP is to find the least upper bound and greatest lower bound which act as approximate solutions.

- The shortest route around a network, travelling along every edge at least once and returning to the starting point is called the **route inspection problem** often called the **Chinese postman problem**.

Algorithms you must know

- **Dijkstra's algorithm** is used to find the shortest path through a network.
  Step 1: assign the permanent label 0 to the starting vertex;
  Step 2: assign temporary labels to all the vertices that are connected directly to the most recently permanently labelled vertex;
  Step 3: choose the vertex with the smallest temporary label and assign a permanent label to that vertex;
  Step 4: repeat steps 2 and 3 until all the vertices have permanent labels;
  Step 5: find the shortest path by tracing back through the network.

- A minimum connector problem involves finding a spanning tree of minimum length. **Kruskal's** and **Prim's algorithms** are used for finding a minimum spanning tree in a network.

- **Prim's algorithm**
  Step 1: choose a starting vertex;
  Step 2: join this vertex to the nearest vertex directly connected to it;
  Step 3: join the nearest vertex, not already in the solution, to any vertex in the solution provided that it does not form a cycle;
  Step 4: repeat until all vertices have been included.

- **Kruskal's algorithm**
  Step 1: rank the edges in order of length;
  Step 2: select the shortest edge in the network;
  Step 3: select from the edges not in the solution the shortest edge which does not form a cycle;
  Step 4: repeat step 3 until all the vertices are in the solution.

- Dijkstra's, Kruskal's and Prim's algorithms are examples of **greedy** algorithms which choose the best option at each stage.

- The basic idea in the **labelling procedure** for solving network flow problems is to find a flow by inspection then to increase its value step by step until the flow cannot be increased any further.

- The **maximum flow/minimum cut algorithm** for the maximum flow through a network states that the maximum flow equals the value of the minimum cut dividing the network.

Don't make these mistakes...

In Dijkstra, don't forget to give temporary labels to all deserving vertices.

Don't confuse Prim and Kruskal!

In Kruskal, don't forget to state the order of arc selection and indicate arcs rejected.

In Prim, don't forget to scan arcs coming from **all** previously selected nodes.

Exam Questions and Student's Answers

How to score full marks

Q1

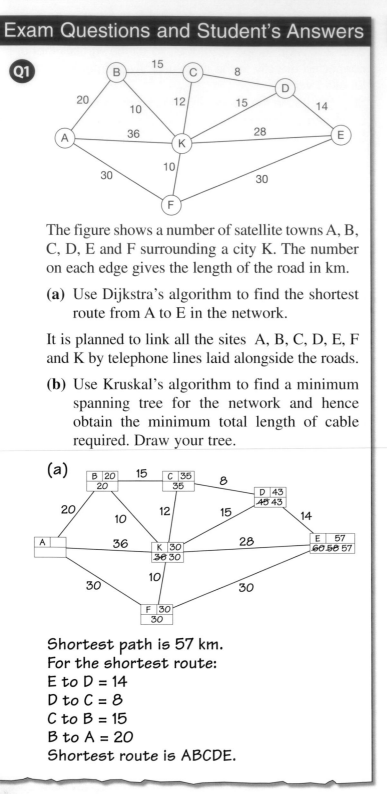

The figure shows a number of satellite towns A, B, C, D, E and F surrounding a city K. The number on each edge gives the length of the road in km.

(a) Use Dijkstra's algorithm to find the shortest route from A to E in the network.

It is planned to link all the sites A, B, C, D, E, F and K by telephone lines laid alongside the roads.

(b) Use Kruskal's algorithm to find a minimum spanning tree for the network and hence obtain the minimum total length of cable required. Draw your tree.

(a)

Shortest path is 57 km.
For the shortest route:
E to D = 14
D to C = 8
C to B = 15
B to A = 20
Shortest route is ABCDE.

- Show your labelling clearly.

- When determining the shortest route, show how the route is obtained by a **backward pass** through the network. Do not just write down the route.

(b) CD – 8, BK – 10, KF – 10, CK – 12,
DE – 14, DK – 15, BC – 15, AB – 20,
EK – 28, AF – 30, FE – 30, AK – 36

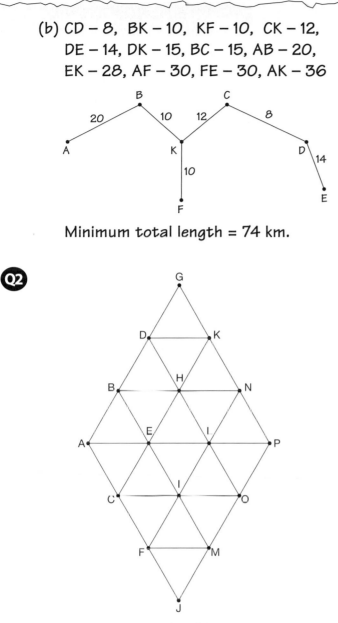

Minimum total length = 74 km.

How to score full marks

● Be careful not to include cycles as you apply **Kruskal's algorithm**. Be careful not to confuse Kruskal with Prim! For Kruskal's algorithm you need to **rank order the edges first** and then select the shortest edges in turn avoiding cycles.

Q2

The network in the figure has 16 vertices.

(a) Given that the length of each edge is 1 unit, find:
  (i) the shortest distance from A to K
  (ii) the length of a minimum spanning tree.

(b) (i) Find the length of an optimal Chinese postman route, starting and finishing at A.
  (ii) For such a route, state the edges that would have to be used twice.
  (iii) Given that the edges AE and LP are now removed, find a new length of an optimal Chinese postman route, starting and finishing at A.

(a) (i) The shortest distance from A to K is 3 km.

    (ii) The minimum spanning tree is 15 km.

(b) (i) Total length of all the edges is 33 km.
Vertices A and P are odd – need to repeat edges – 3 km.
Total length of route is 36 km.

    (ii) Repeat AE, EL, LP.

    (iii) Vertices E and L are odd – repeat edge EL.
Length of the optimal route is 31 + 1 = 32 km.

How to score full marks

- There are several routes from A to K with length 3 km.

- It is sufficient to write down the answer.

- Remember to look for any **odd vertices** so that you can look for the need for repeated edges.

Q3 The island of Lanzarote is short of water and the local government is to implement a new system so that water is distributed to all the main towns on the island. The main source of water is at Arrecife (A). The figure shows the main towns and the distances, in kilometres, between them.

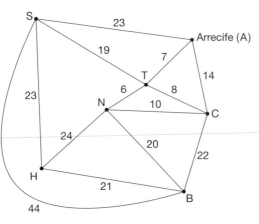

(a) Using Prim's algorithm starting at A, and showing your working at each stage, find the minimum length of piping needed. Show your final network of piping as a minimum spanning tree.

(b) A tourist lands at the airport at Arrecife (A) and is to tour the island, visiting each of the other six towns before returning to A.

    (i) By deleting A, find a lower bound for the length of such a tour.

    (ii) Explain why a tour of such length is impossible in this situation.

(a)

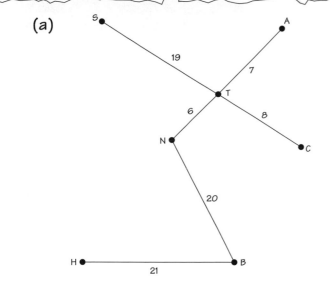

Minimum length is 81 km.

(b) (i) Delete A – minimum spanning tree
is 74 km.
Lower bound is 74 + 2 × 7 = 88 km.
(ii) There is not a complete tour
because the vertices (nodes) S, C
and H are single nodes from which
we would need to return.

Q4

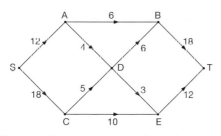

The figure shows a capacitated network. The numbers on each arc indicate the capacity of that arc in appropriate units.

(a) Explain why it is not possible to achieve a flow of 30 through the network from S to T.

(b) State the maximum flow along
(i) SABT, **(ii)** SCET

(c) Show these flows on a diagram of the network.

(d) Taking your answer to part (c) as the initial flow pattern, use the labelling procedure to find a maximum flow from S to T. List each flow-augmenting path you use together with its flow.

(e) Indicate a maximum flow on your diagram.

(f) Prove that your flow is maximal.

How to score full marks

● Be careful not to confuse Prim with Kruskal! For **Prim's algorithm** you **choose the start vertex (node) and join the nearest vertex to it.** Join the next nearest not in the solution to any vertex in the solution. Remember to avoid cycles.

(a) Flow out of A is 10 so max flow along
SA is 10 – not 12.
Flow out of $S \leq 28 < 30$

(b) (i) SABT – 6
(ii) SCET – 10

(c)

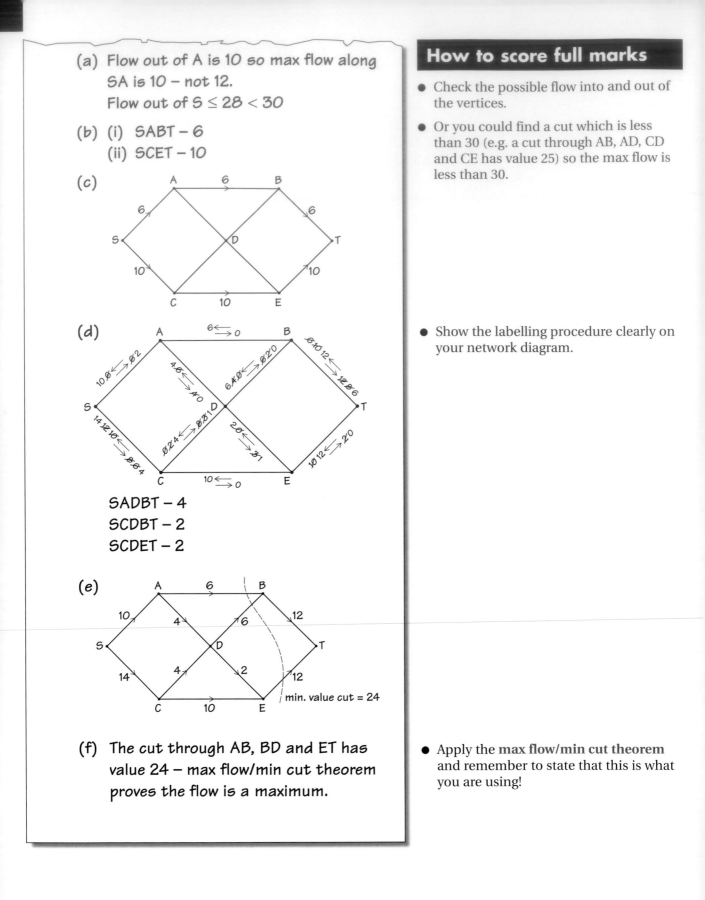

- Check the possible flow into and out of the vertices.

- Or you could find a cut which is less than 30 (e.g. a cut through AB, AD, CD and CE has value 25) so the max flow is less than 30.

(d)

SADBT – 4
SCDBT – 2
SCDET – 2

- Show the labelling procedure clearly on your network diagram.

(e)

min. value cut = 24

(f) The cut through AB, BD and ET has value 24 – max flow/min cut theorem proves the flow is a maximum.

- Apply the **max flow/min cut theorem** and remember to state that this is what you are using!

Questions to try

Q1 **(a)** Explain why it is impossible to draw a network with exactly three odd vertices.

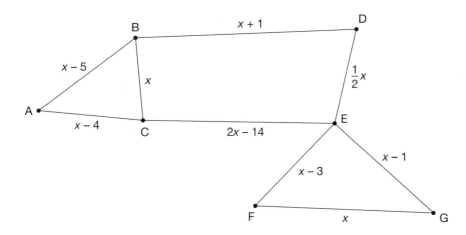

The Route Inspection problem is solved for the network shown in the figure and the length of the route is found to be 100.

(b) Determine the value of x, showing your working clearly.

Q2 **(a)** Explain the main differences between Prim's and Kruskal's algorithms for solving minimum connector problems.

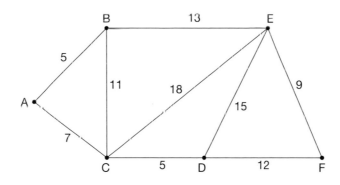

(b) Listing the edges (arcs) in the order that you select them, find a minimum connector for the network shown in the figure, using

    (i) Prim's algorithm

    (ii) Kruskal's algorithm.

Q3 Use Dijkstra's algorithm to find the shortest path from A to F in the following network.

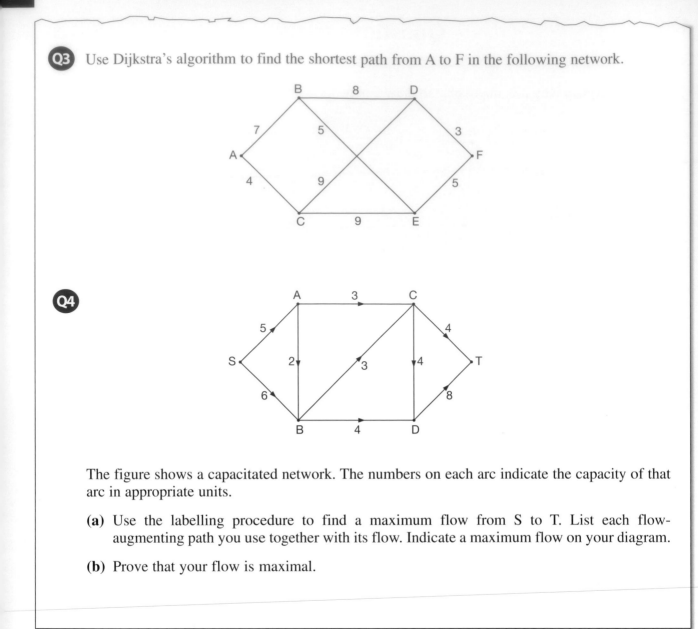

The figure shows a capacitated network. The numbers on each arc indicate the capacity of that arc in appropriate units.

(a) Use the labelling procedure to find a maximum flow from S to T. List each flow-augmenting path you use together with its flow. Indicate a maximum flow on your diagram.

(b) Prove that your flow is maximal.

Answers can be found on pages 149–150.

Key points to remember

- When we are planning a project we usually want to know the minimum time that is needed to complete the project and the order of the various tasks that make up the project.

- The **critical activities** in the project are those activities that must be started on time to avoid delaying the whole project.

- The **critical path** is the minimum time needed to complete a project, following the longest path through the precedence network and including all of the critical activities.

Algorithms you must know

The algorithm for finding a critical path consists of four main steps:

- Draw a precedence network to show the necessary steps in a project;

- Find the critical path through the network;

- Draw cascade charts showing precedence relations, to make the best use of the resources available;

- Use resource levelling to balance the amount of work to be done and the resources available for a project.

Don't make these mistakes...

Do at least one rough drawing first!

Don't forget arrows.

Don't have more than one start point and finish point.

Don't use too many dummies.

On a scheduling diagram, make sure every activity is present only once.

Q1 A conservatory is to be built on the back of a house. Details of the jobs to be done and timings are shown below:

	Activity	Duration (days)	Preceding activity
A	Lay the foundations	3	
B	Build the walls	5	C
C	Lay the drains	2	A
D	Lay the floor	1	B
E	Make the door and window frames	2	
F	Erect the roof	3	D
G	Fit the door frame and door	0.5	E, F
H	Fit the windows	2	G
I	Plaster the walls and ceiling	8	H

Draw a precedence network to show this information.

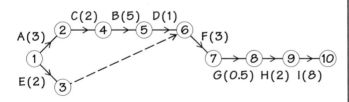

- **Label the network clearly** showing the activity and the duration. Include arrows on each edge (arc). Remember to include any dummy edges (arcs) as dashed lines.

Q2

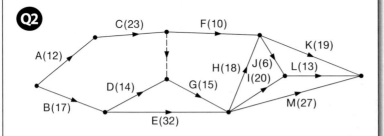

The network in the figure shows the activities involved in the process of producing a perfume. The activities are represented by the arcs. The number in brackets on each arc gives the time, in hours, taken to complete the activity.

(a) Calculate the early time and the late time for each event.

(b) Hence determine the critical activities.

(c) Calculate the total float time for D.

Each activity requires only one person.

(d) Find a lower bound for the number of workers needed to complete the process in the minimum time.

Given that there are only three workers available, and that the workers may **not** share an activity,

(e) Schedule the activities so that the process is completed in the shortest time. State the new shortest time.

(a)

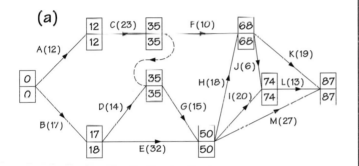

- Don't make silly arithmetical mistakes.

(b) A, C, G, H, J, K, L

(c) 35 − 17 − 14 = 4

- Show your working for the float time.

(d) Total hours needed for all the tasks is 226 so we need a minimum of $\dfrac{226}{87}$ = 2.6 i.e. 3 workers.

(e)

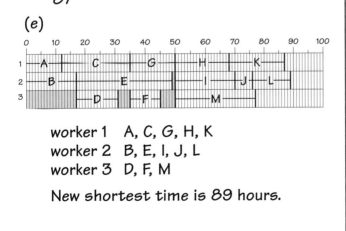

- There may be other ways of allocating the tasks to the workers.

worker 1   A, C, G, H, K
worker 2   B, E, I, J, L
worker 3   D, F, M

New shortest time is 89 hours.

Q1 **(a)** Find the early and late times for each activity in the precedence diagram in the figure.

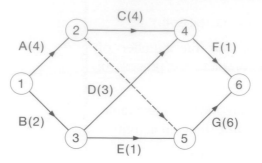

(b) List the critical activities.

(c) Give the total float for activity C.

Q2 A project is modelled by the activity network shown in the figure below. The activities are represented by the arcs. The number in brackets on each arc gives the time, in hours, taken to complete the activity. The left box entry at each vertex is the earliest event time and the right box entry is the latest event time.

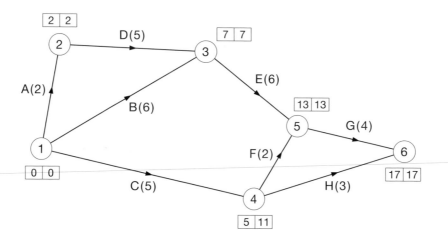

(a) Determine the critical activities and the length of the critical path.

(b) Obtain the total floats for the non-critical activities.

(c) Draw a cascade (Gantt) chart showing the information found in parts (a) and (b).

Given that each activity requires one worker,

(d) Draw up a schedule to determine the minimum number of workers required to complete the project in the critical time. State the minimum number of workers.

Answers can be found on page 150.

Key points to remember

- If there are several solutions to a problem then the **optimum solution** is the best solution according to some given conditions.

- The **matching problem** consists of matching two subsets in an optimal way. A bipartite graph is used to model the lists in a matching problem.

- In a **linear programming problem** we find the maximum or minimum of a linear function (called the objective function) which is subject to a set of constraints which are linear inequalities or equalities.

- To solve a linear programming problem we first formulate the objective function and the constraints. For problems involving two variables the constraints form lines in the x–y plane and the problem is solved graphically. For problems with three or more variables we use the simplex method.

Algorithms you must know

- The optimal solution to the matching problem is called the **maximum matching**. The algorithm for finding the maximum matching consists of two parts: first the labelling procedure which assigns each vertex with a label using an initial matching and second, the labels are used to find an alternating path. An improved matching then consists of appropriate edges from the initial matching and the alternating path.

- The **simplex method** involves three main steps:

  Step 1: introduce slack variables so that inequalities become equalities;

  Step 2: set up and manipulate the simplex tableau using the Gaussian elimination manipulations until an optimal solution is reached;

  Step 3: interpret the solution in the context of the problem.

Don't make these mistakes ...

Don't forget to **interpret the solution** in the context of the original problem.

Don't forget the non-negativity contraints, $x \geq 0$, etc.

Don't forget to draw the lines accurately; use a ruler with a sharp pencil!

Don't forget to show the feasible region clearly.

Q1 Six workers A, B, C, D, E and F are to be matched to six tasks 1, 2, 3, 4, 5 and 6.

The table below shows the tasks that each worker is able to do.

Worker	Tasks
A	2, 3, 5
B	1, 3, 4, 5
C	2
D	3, 6
E	2, 4, 5
F	1

A bipartite graph showing this information is drawn below.

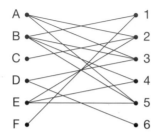

Initially, A, B, D and E are allocated to tasks 2, 1, 3 and 5 respectively.

Starting from the given initial matching, use the matching improvement (maximum matching) algorithm to find a complete matching, showing your alternating paths clearly.

The initial matching:

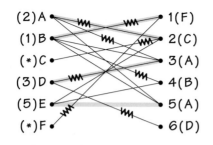

Alternating paths are
4–B–1–F and 6–D–3–A–2–C.

Edge not in the alternating paths is E–5.

Edges in the alternating paths but not in the initial matching are A–3, B–4, C–2, D–6, F–1.

A complete matching is A–3, B–4, C–2, D–6, E–5, F–1.

- Draw the given initial matching M.
- Label carefully each vertex using the labelling procedure.
- Mark on your graph the alternating paths, remember to start with a vertex on the right that is not in the initial matching and follow the labels back to a (*) label.
- Identify the edges in each alternating path that are in the initial matching but are not in the alternating paths.
- Identify the edges in each alternating path that are not in the initial matching but are in the alternating paths.
- An improved matching consists of this set of edges. In this question it is a complete matching so we can stop. However in other questions we may need to repeat the algorithm using our improved matching as a new initial matching. In these questions the solution is not necessarily unique.

Q2 A company produces two types of self-assembly wooden bedroom suites, the 'Oxford' and the 'York'. After the pieces of wood have been cut and finished, all the materials have to be packaged. The table below shows the time, in hours, needed to complete each stage of the process and the profit made, in pounds, on each type of suite.

	Oxford	York
Cutting	4	6
Finishing	3.5	4
Packaging	2	4
Profit (£)	300	500

The times available each week for cutting, finishing and packaging are 66, 56 and 40 hours respectively.

The company wishes to maximise its profit.

Let x be the number of Oxford, and y be the number of York suites made each week.

(a) Write down the objective function.

(b) In addition to

$2x + 3y \leq 33,$
$x \geq 0,$
$y \geq 0,$

find two further inequalities to model the company's situation.

(c) On a graph, illustrate all the inequalities, indicating clearly the feasible region.

(d) Explain how you would locate the optimal point.

(e) Determine the number of Oxford and York suites that should be made each week and the maximum profit gained.

It is noticed that when the optimal solution is adopted, the time needed for one of the three stages of the process is less than that available.

(f) Identify this stage and state how many hours the time may be reduced.

(a) $P = 300x + 500y$

(b) $3.5x + 4y \leq 56$
$2x + 4y \leq 40$

(c)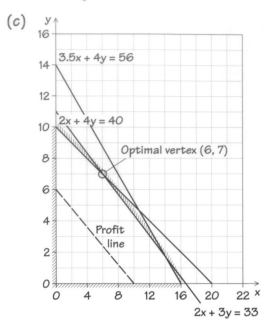

How to score full marks

- Label the graph of each constraint line and show the feasible region clearly on your graph.

(d) Draw in the profit line and select the point on the profit line furthest from the origin.

- Alternatively you could find the coordinates of each corner of the feasible region, find the profit at each corner and select the maximum.

(e) $x = 6$ and $y = 7$
The number of Oxford suites is 6 and the number of York suites is 7.
The maximum profit is £5300.

- Remember to **give the answer in words**.

(f) The line $3.5x + 4y = 49$ passes through the optimum vertex (6, 7) so the time needed for finishing can be reduced by 7 hours.

Q3 The tableau below is the initial tableau for a maximising linear programming problem.

Basic variable	x	y	z	r	s	Value
r	2	3	4	1	0	8
s	3	3	1	0	1	10
P	−8	−9	−5	0	0	0

(a) For this problem $x \geq 0$, $y \geq 0$, $z \geq 0$. Write down the other two inequalities and the objective function.

(b) Solve this linear programming problem.

(c) State the final value of P, the objective function, and each of the variables.

(a) $2x + 3y + 4z \leq 8$

$3x + 3y + z \leq 10$

$P = 8x + 9y + 5z$

(b)

Basic variable	x	y	z	r	s	Value
r	2	③	4	1	0	8
s	3	3	1	0	1	10
P	−8	−9	−5	0	0	0

Basic variable	x	y	z	r	s	Value	
y	$\frac{2}{3}$	1	$\frac{4}{3}$	$\frac{1}{3}$	0	$\frac{8}{3}$	R1 ÷ 3
s	①	0	−3	−1	1	2	R2 − 3R1
P	−2	0	7	3	0	24	R3 + 9R1

Basic variable	x	y	z	r	s	Value	
y	0	1	$\frac{10}{3}$	1	$-\frac{2}{3}$	$\frac{4}{3}$	R1 − $\frac{2}{3}$R2
x	1	0	−3	−1	1	2	
P	0	0	1	1	2	28	R3 + 2R2

(c) $P = 28$

$x = 2$, $y = \dfrac{4}{3}$, $z = 0$, $r = 0$ and $s = 0$.

- **Identify** the **pivot column** and **pivot element**.
- **Show your row operations** clearly.

- Remember to give the **values of the slack variables**.

Questions to try

Q1 A youth club wishes to enter a team at the regional athletics meeting. They have six good runners, A, B, C, D, E and F, who can take part and there are six track events. Runner A can run the 100 m and 200 m, runner B can run the 800 m and 1500 m, runner C can run the 200 m and 100 m hurdles, runner D can run the 400 m, 800 m and 1500 m, runner E can run the 400 m and 800 m and runner F can run the 100 m and 100 m hurdles. How should the captain pick the team?

Q2 The Tony television company makes analogue and digital televisions. Both types of television require a number of component A and component B.

Each analogue television requires 2 of component A and 3 of component B.
Each digital television requires 4 of component A and 1 of component B.

Each day:
   the company has 50 of component A and 24 of component B; and the company is to make at least 2 of each type of television, but no more than 20 in total.

The company sells each analogue television at a profit of £20 and each digital television at a profit of £25.
Each day, the company makes and sells x analogue and y digital televisions.
The company needs to find its minimum and maximum daily income, £T.

(a) Formulate the company's situation as a linear programming problem.

(b) Draw a suitable diagram to enable the problem to be solved graphically, indicating the feasible region and the direction of the objective line.

(c) Use your diagram to find the company's minimum and maximum daily income, £T.

Q3 Use the simplex tableau to solve the following linear programming problem.

Maximise $f = 9x + 4y$
where $\quad 3x + 4y \leq 48$
$\quad\quad 2x + y \leq 17$
$\quad\quad 3x + y \leq 24$
$\quad\quad x \geq 0, y \geq 0$

Answers can be found on pages 151–152.

26 Answers to Questions to try

1 Algebra and equations

Answers | **How to solve these questions**

Q1

(a) $a^5 \times a^{-2} = a^{5-2} = a^3$

(b) $\dfrac{a^5}{a^{-3}} = a^{5-(-3)} = a^8$

(c) $\left(\dfrac{\sqrt[4]{a}}{\sqrt[7]{a}}\right)^3 = (a^{\frac{1}{4}-\frac{1}{7}})^3$

$= (a^{\frac{3}{28}})^3$

$= a^{\frac{9}{28}}$

How to solve:

$a^m \times a^n = a^{m+n}$

$(a^m)^n = a^{mn}$

Q2 For two distinct real roots:

$k^2 - 36 > 0$

$\Rightarrow k < -6 \text{ or } k > 6 \text{ i.e. } |k| > 6.$

(i) $k = -4\sqrt{3}$

$\Rightarrow f(x) = x^2 - 4\sqrt{3}x + 9$

$\Rightarrow f(x) = (x - 2\sqrt{3})^2 - 3$

$a = -2\sqrt{3}$ and $b = -3$.

The least value of f(x) is –3 (occurs when $x = 2\sqrt{3}$).

How to solve: First, substitute for the given value of k. Then to complete the square, the value of a is half the coefficient of x.

(ii) $f(x) = 0$

$\Rightarrow (x - 2\sqrt{3})^2 - 3 = 0$

$x = 2\sqrt{3} \pm \sqrt{3}$

i.e. $x = 3\sqrt{3} \text{ or } \sqrt{3}$.

How to solve: You do not need to use the formula here since (i) gives the answer easily.

Q3 $x^2 > x + 20$

$\Rightarrow x^2 - x - 20 > 0$

$\Rightarrow (x - 5)(x + 4) > 0$

$\Rightarrow x < -4 \text{ or } x > 5$

How to solve: You need to factorise and remember that f(x) > 0 if the factors are both positive or both negative.

Q4 $x + x + 1 > 10$

$2x + 1 > 10$

$x > 4.5$

$x(x + 1) < 72$

$x^2 + x - 72 < 0$

$(x - 8)(x + 9) < 0$

$-9 < x < 8$

So the numbers can be 5 and 6, 6 and 7 or 7 and 8.

How to solve: Let the numbers be x and $x + 1$. Form and solve an equality by considering the sum of the two numbers.

Form and solve an equality by considering the product of the two numbers.

Remember that you are looking for pairs of consecutive numbers.

Answers | **How to solve these questions**

Q5

How to solve: You should always draw a diagram and mark in the side lengths.

(a) $x + x - 5 + x + x - 5 > 32$

$= 4x - 10 > 32$

$= 4x > 42$

$= x > 10.5$

(b) $x(x-5) < 104$

(c) $x^2 - 5x - 104 < 0$ from (b)

$\Rightarrow (x-13)(x+8) < 0$

$\Rightarrow -8 < x < 13$

Since $x > 10.5$ from (a), then $10.5 < x < 13$

How to solve: Form a quadratic by equating the two expressions for y.

Q6

(i) $x^2 - 3x + 2 = 3x - 7$

$x^2 - 6x + 9 = 0$

$(x - 3)^2 = 0 \Rightarrow x = 3$

How to solve: Since there is only one solution this means that the line touches the parabola once, i.e. it is a tangent.

(ii) $y = 3x - 7$ is a tangent to $y = x^2 - 3x + 2$.

How to solve: The discriminant of $ax^2 + bx + c = 0$ is $b^2 - 4ac$

Q7

(a) Discriminant

$= k^2 - 4 \times 1 \times k = k^2 - 4k$

(b) $k^2 - 4k < 0$

$k(k - 4) < 0$

$0 < k < 4$

How to solve: For there to be no real roots the discriminant must be less than zero.

Q8

(a) $x^2 - 4 \times 1 \times 16 = 0$

$x^2 = 64$

$m = \pm 8$

How to solve: Using the standard $ax^2 + bx + c = 0$ notation, the solutions depend on the value of $b^2 - 4ac$, which is often called the discriminant. This is the part of the quadratic formula that is square rooted.

For equal roots $b^2 - 4ac = 0$.

(b) $x^2 - 4 \times 1 \times 16 > 0$

$x^2 > 64$

$m > 8 \text{ or } m < -8$

For two distinct roots $b^2 - 4ac > 0$.

(c) $x^2 - 4 \times 1 \times 16 < 0$

$x^2 < 64$

$-8 < m < 8$

For no real roots $b^2 - 4ac < 0$.

Answers

Q9

(a) $x^2 - 8x + 3 = (x-4)^2 - 16 + 3$
$= (x-4)^2 - 13$

(b) Minimum value of y is -13 when $x = 4$, so the coordinates are $(4, -13)$.

Q10

(a) $\dfrac{\sqrt{2}+1}{\sqrt{2}-1} = \dfrac{(\sqrt{2}+1)(\sqrt{2}+1)}{(\sqrt{2}-1)(\sqrt{2}+1)}$
$= \dfrac{2 + 2\sqrt{2} + 1}{2 - 1}$
$= 2\sqrt{2} + 3$

(b) $\sqrt{2}(x - \sqrt{2}) < x + 2\sqrt{2}$
$x\sqrt{2} - 2 < x + 2\sqrt{2}$
$x\sqrt{2} - x < 2\sqrt{2} + 2$
$x(\sqrt{2} - 1) < 2(\sqrt{2} + 1)$
$x < \dfrac{2(\sqrt{2}+1)}{(\sqrt{2}-1)}$
$x < 4\sqrt{2} + 6$

2 ARITHMETIC AND GEOMETRIC PROGRESSIONS

Q1 $a = 7$, $d = 2$

$u_{18} = 7 + (18-1) \times 2 = 41$

$S_{15} = \dfrac{1}{2} \times 15 \times (2 \times 7 + (15-1) \times 2)$
$= 315$

Q2 $a = 100$, $r = 0.9$

(a) $u_{10} = 100 \times 0.9^9$
$= 38.742$

(b) $S_{20} = 100 \left(\dfrac{1 - 0.9^{20}}{1 - 0.9} \right)$
$= 878.4$ (1 d.p.)

(c) $S_\infty = \dfrac{100}{1 - 0.9}$
$= 1000$

How to solve these questions

Q9

(a) Here you have to complete the square.

(b) The minimum value of y is obtained when the bracket is zero.

Q10

(a) Here the denominator needs to be rationalised. This is done by multiplying both the numerator and the denominator of the fraction by $(\sqrt{2} - 1)$.

(b) First expand the brackets.

Then bring all the terms with x to the left hand side and factorise.

Note that as $(\sqrt{2} - 1)$ is positive, the inequality remains unchanged as a result of the division. The final result can be simplified using the result from part (a).

2 ARITHMETIC AND GEOMETRIC PROGRESSIONS

Q1 First note that this is an AP. Identify the first term and the common difference.

Substitute $n = 18$, $a = 7$ and $d = 2$ into the formula for the nth term of an AP.

Substitute $n = 15$, $a = 7$ and $d = 2$ into the formula for the sum of n terms of an AP.

Q2 You should notice that this is a GP with first term 100 and common ratio 0.9.

(a) Use the formula for the nth term of a GP, with $n = 10$.

(b) Use the formula for the sum of n terms of a GP with $n = 20$.

(c) As $|r| < 1$, you can use the formula for the sum to infinity.

Answers

Q3

(a) $S_{20} = \dfrac{1}{2} \times 20(2 \times 4 + (20-1) \times 5)$
$= 1030$

(b) $4 + 5(n-1) > 131$
$5n - 1 > 131$
$n > 26.4$
∴ Smallest n is **27**.

Q4 $a + 7d = 40$
$a + 19d = 124$
∴ $d = 7$ and $a = -9$
$S_{20} = \dfrac{20(2 \times -9) + (20 - 1 \times 7)}{2}$
$= 1150$

Q5

(a) $ar = 80$
$ar^4 = 5.12$
∴ $r = \dfrac{2}{5}$ and $a = 200$

(b) $S_\infty = \dfrac{200}{1 - \frac{2}{5}} = \dfrac{1000}{3}$

(c) $S_{14} = 200 \left(\dfrac{1 - \left(\frac{2}{5}\right)^{14}}{1 - \left(\frac{2}{5}\right)} \right)$
$= 333.332\,44$
$S_\infty - S_{14} = 8.9 \times 10^{-4}$

Q6 $ar = 1$
$\dfrac{a}{1 - r} = 7.2$
$7.2r^2 - 7.2r + 1 = 0$
∴ $r = \dfrac{1}{6}$ or $r = \dfrac{5}{6}$

For $r = \dfrac{5}{6}$, $a = \dfrac{6}{5}$ and the sequence begins

$\dfrac{6}{5}, 1, \dfrac{5}{6}, \dfrac{25}{36}, \cdots$

How to solve these questions

Q3

(a) Use the formula for the sum of n terms of an AP with $n = 20$, $a = 4$ and $d = 5$.

(b) Form and solve an inequality based on the formula for the nth term of an AP, in this case $u_n > 131$.

Q4 By using the formula for the nth term of an AP you can form a pair of simultaneous equations.

Solve them to find d and a.

Once you know d and a you can use the formula for the sum of n terms with $n = 20$ to find the required sum.

Q5 First, use the formula for the nth term of a GP to form a pair of simultaneous equations.

Solve them to find r and a.

(b) As $|r| < 1$, you can use the formula for the sum to infinity.

(c) First use the formula for the sum of the first n terms of a GP to find the sum of the first 14 terms.

Then find the difference between the two sums.

Q6 You can form this equation because the second term is 1.

You can form this equation as the sum to infinity is 7.2.

Substitute $a = \dfrac{1}{r}$ in the second equation and simplify to find this quadratic equation.

Then you can solve the quadratic to obtain these solutions.

Use the larger value of r to find a.

$$S_{20} = \frac{6}{5}\left(\frac{1-\left(\frac{5}{6}\right)^{20}}{1-\frac{5}{6}}\right)$$

$$= 7.012 \text{ (3 d.p.)}$$

Q7 $1932 = 1200 \times \left(1+\frac{r}{100}\right)^4$

$1 + \frac{r}{100} = \sqrt[4]{\frac{1932}{1200}}$

$r = 12.64$ (correct to 2 d.p.)

$$S_{10} = 1200\left(\frac{1-\left(1+\frac{r}{100}\right)^{10}}{1-\left(1+\frac{r}{100}\right)}\right)$$

$= £21\,724$

Q8 A: $S_{10} = \frac{1}{2} \times 10 \times$

$(2 \times 1000 + (10-1) \times 100)$

$= £14\,500$

B: $14\,500 = X\left(\frac{1-1.1^{10}}{1-1.1}\right)$

$X = £909.81$

Q9 $d_2 = 2 + \frac{2}{5} \times 2 + \frac{2}{5} \times 2$

$= 3.6$ m

$d_\infty = 2 + \frac{2}{5}\times 2 + \frac{2}{5}\times 2 + \left(\frac{2}{5}\right)^2$

$\times 2 + \left(\frac{2}{5}\right)^2 \times 2 + \dots$

$= 2 + 2 \times \left(\frac{\frac{2}{5}\times 2}{1-\frac{2}{5}}\right)$

$= 2 + 2 \times \frac{4}{3}$

$= \frac{14}{3}$

Then use the formula for the sum of n terms of a GP.

Q7 You can use the fact that the 5th term is 1932 to form this equation.

Solve the equation to find r.

Then use the formula for the sum of n terms to find the total that is paid into the fund. Note that this answer is obtained using the exact value of r.

Q8 Use the formula for the sum of an AP to find the total for scheme A.

Form and solve an equation based on a GP with 10 terms and sum 14 500, using the formula for the sum of n terms of a GP.

Q9 The total distance is made up of a fall of 2 m, a rise of $\frac{2}{5} \times 2$ and a fall of the same distance.

The total distance is given by this expression which is 2 plus twice a GP with first term $\frac{2}{5} \times 2$ and common ratio $\frac{2}{5}$. You can sum the GPs using the formula for the sum to infinity.

Q10 $r = \frac{2400}{3000} = 0.8$

(i) $u_{20} = 3000 \times 0.8^{20-1} = 3000 \times 0.18^{19}$

$= 43$

(ii) $S_{20} = 3000\left(\frac{1-0.8^{20}}{1-0.8}\right) = 14827$

(iii) $S_\infty = \frac{3000}{1-0.8} = 15000$

3 TRIGONOMETRY

Q1 $2r\theta = 5r$

$\theta = 2.5$

Area $= \frac{1}{2} \times (2r)^2 \times 2.5 = 5r^2 \text{ cm}^2$

Q2 Area of sector

$= \frac{1}{2} \times 6^2 \times 0.7 = 12.6 \text{ cm}^2$

Area of triangle

$= \frac{1}{2} \times 6 \times 6\sin 0.7$

$= 11.60 \text{ cm}^2$

Area shaded $= 12.60 - 11.60$

$= 1.00 \text{ cm}^2$

Perimeter

$= 6 \times 0.7 + 2 \times 6\sin 0.35$

$= 8.31 \text{ cm}$

Q3

(a) $\frac{\sin C}{3} = \frac{\sin 2.1}{5}$

$C = \sin^{-1}\left(\frac{3\sin 2.1}{5}\right) = 0.544$

(b) $A = \pi - 2.1 - 0.5444 = 0.4972$

Area $= \frac{1}{2} \times 3 \times 5 \sin 0.4972$

$= 3.58 \text{ cm}^2$

(c) Perimeter $= 3 + 3 + 3 \times 0.4972$

$= 7.49 \text{ cm}$

Area $= \frac{1}{2} \times 3^2 \times 0.4972 = 2.24 \text{ cm}^2$

Q10 First calculate the common ratio, in this case using $r = \frac{u_2}{u_1}$.

This can then be used in the rest of the question.

(i) The 20th term of the GP will give the sales in the 20th week and can be calculated using the formula for the nth term.

(ii) The total sales in the first 20 weeks is given by the sum of the first 20 terms, which can be calculated using the formula for the sum of n terms of a GP.

(iii) The total sales can be found using the formula for the sum to infinity.

Q1 Use the formula for arc length to find θ. Then you can use this value of θ to find the area.

Q2 You should first find the area of the sector.

When you find the area of the triangle the height is given by 6 sin0.7.

Q3

(a) Use the sine rule to find the angle, being careful to make sure that you do work in radians.

(b) First calculate the angle A and then use the formula for the area of a triangle.

(c) The perimeter is made up of twice the radius plus the arc length.

The standard formula can be used to find the area of the sector.

Answers

Q4 $\sin^{-1}(-0.5) = -30°$
$\theta = 180° + 30° = \mathbf{210°}$ or
$\theta = 360° - 30° = \mathbf{330°}$

Q5 $\sin 2x = \dfrac{1}{2}$
$2x = 30°, 150°, 390°$ or $510°$
$x = \mathbf{15°, 75°, 195°}$ or $\mathbf{255°}$

Q6 $\dfrac{\sin\theta}{\cos\theta} = 2\sin\theta$
$\sin\theta = 2\sin\theta\cos\theta$
$2\sin\theta\cos\theta - \sin\theta = 0$
$\sin\theta = 0$ or $\cos\theta = \dfrac{1}{2}$
$\theta = \mathbf{60°, 180°}$ or $\mathbf{300°}$

Q7 $2(1 - \cos^2 x) + 3\cos x = 0$
$2\cos^2 x - 3\cos x - 2 = 0$
$\cos x = -0.5$ or 2
$\cos^{-1}(-0.5) = \mathbf{120°}$
or $x = 360° - 120°$
$= \mathbf{240°}$

Q8 $15(1 - \sin^2\theta) = 13 + \sin\theta$
$15\sin^2\theta + \sin\theta - 2 = 0$
$\sin\theta = -\dfrac{2}{5}$ or $\dfrac{1}{3}$
$\theta = \mathbf{19.5°, 160.5°, 203.6°}$
or $\mathbf{336.4°}$

Q9
(a) $15(1 - \cos^2\theta) = 13 + \cos\theta$
$15 - 15\cos^2\theta = 13 + \cos\theta$
$\mathbf{15\cos^2\theta + \cos\theta - 2 = 0}$

(b) $(3\cos\theta - 1)(5\cos\theta + 2) = 0$
$\cos\theta = \dfrac{1}{3}$ or $\cos\theta = -\dfrac{2}{5}$
$\theta = 70.5°$
$\theta = 113.6°$
or
$\theta = 360° - 70.5° = 289.5°$
$\theta = 360° - 113.6° = 246.4°$

How to solve these questions

Q4 You can obtain a first solution from your calculator. You can then calculate the other solutions, using a sketch graph if necessary.

Q5 You should first find the values of $2x$ that satisfy the equation.
The first of these is $\sin^{-1}\dfrac{1}{2} = 30°$.
Then you can calculate the others, but you must include solutions in the range $0 \le 2x \le 720°$. Finally, divide these solutions by 2 to obtain the values of x.

Q6 First, use $\tan\theta = \dfrac{\sin\theta}{\cos\theta}$.
You can then find the solutions to this equation.

Q7 First, replace $\sin^2 x$ by $1 - \cos^2 x$.
This will give you a quadratic equation in $\cos x$.
You can solve this to give these two solutions. You find the first solution from your calculator. Subtracting this from 360° will give the second solution.

Q8 First, replace $\cos^2\theta$ by $1 - \sin^2\theta$.
This gives you a quadratic equation in $\sin\theta$.
You find two solutions when you solve this.
You can then find the values of θ.

Q9
(a) Use $\sin^2\theta = 1 - \cos^2\theta$ to rearrange the equation.

(b) In this case the quadratic has been factorised, but you can use any method to do this. Remember to find all of the solutions of the trigonometric equations in the range specified.

Answers

Q10
(a) (i) $8 - 2 = \mathbf{6\text{ m}}$
(ii) $8 + 2 = \mathbf{10\text{ m}}$

(b) $2\pi = 12.4k$
$k = \dfrac{2\pi}{12.4}$
$= \mathbf{0.507}$ (3 d.p.)

(c)

4 COORDINATE GEOMETRY

Q1 Gradient of AB $= \left(\dfrac{14 - 6}{5 - 1}\right) = 2$
Equation of p is $y = 2x + c$.
$x = 1$ and $y = 6$ gives $c = 4$
Equation of p is $y = 2x + 4$.
The midpoint has
coordinates $\left(\dfrac{1 + 5}{2}, \dfrac{6 + 14}{2}\right)$
$= (3, 10)$
Equation of q is $y = -\dfrac{1}{2}x + c$
$x = 3$ and $y = 10$ gives
$c = \dfrac{23}{2}$

The equation is $y = -\dfrac{1}{2}x + \dfrac{23}{2}$.

Q2
(a) $3(3x - 6) + x - 12 = 0$
$10x = 30$
$x = 3$
The coordinates are $\mathbf{(3, 3)}$.

How to solve these questions

Q10 For this you take $\cos kt = 1$.
For this you take $\cos kt = -1$.

You should note that a period is 12.4 hours.

You can then form and solve an equation for k.

You should show the minimum and maximum values, and that the period is 12.4.

Q1 First you need to find the gradient of the line through the two points.

You can then substitute the gradient into the equation of a straight line.

You can use the coordinates of A to find the value of c.

Then substitute the value of c in the equation.

The next step is to find the coordinates of the midpoint of AB.

As the lines are perpendicular, the gradient of $q = \dfrac{-1}{\text{gradient of } p} = -\dfrac{1}{2}$ and so you can substitute this value into the equation of a straight line.

You can then use the coordinates of the midpoint to find the value of c.

Finally, substitute the value of c into the equation.

Q2 From the first equation $y = 3x - 6$.

Substitute this into the second equation and solve for x.

Substitute $x = 3$ into either equation to find the y-coordinate.

(b) Gradients are 3 and $-\frac{1}{3}$, so their product is -1, hence the lines are **perpendicular.**

(c) The corners of the triangle are (3, 3), (2, 0) and (12, 0).
Area $= \frac{1}{2} \times (12-2) \times 3$
$= 15$ square units

Q3 The equation of p is
$y = \frac{1}{3}x + c.$

$x = -2$ and $y = 0$ gives $c = \frac{2}{3}$

$y = \frac{1}{3}x + \frac{2}{3}$ or $3y - x - 2 = 0$

The equation of q is
$y = -3x + c.$

$x = 8$ and $y = 0$ gives $c = 24.$

$y = -3x + 24$ or
$y + 3x - 24 = 0$

Q4
(a) The gradient of AC is
$\left(\frac{2-6}{7-3}\right) = -1.$

The equation is $y = -x + c.$
$x = 3$ and $y = 6$ gives $c = 9.$
$y = -x + 9$ or $y + x - 9 = 0$

The gradient of q is 1, hence the equation is $y = x + c.$
$x = 6$ and $y = 5$ gives $c = -1.$
$y = x - 1$ or $y - x + 1 = 0$

(b) The lines intersect when
$x - 1 = -x + 9$ or $x = 5$
The coordinates are (5, 4).

How to solve these questions

Find the gradients by writing the equations in the form $y = mx + c$ and then find their product.

As the lines intersect the x-axis you can find the coordinates of the other corners by substituting $y = 0$ into the equations.

You can now find the area. A sketch of the triangle may help you.

First, note that the gradient of the line will be 3 and substitute this in the equation of a straight line.

You can use the coordinates of the point that has been given to find c.

As the line q is perpendicular to p, you can deduce that its gradient will be 3.

The coordinates of the point on the line can be used to find c.

Find the coordinates of the point of intersection by equating the two equations and solving for x.

Substitute $x = 7$ into either equation to find the y-coordinate.

First, use the coordinates to calculate the gradient of the line AC.

Then write down the equation.

Now use one set of coordinates to find c and complete the equation.

Find the gradient, by using the fact that the product of the two gradients is -1.

This allows you to write down the equation.

You can use the coordinates of B to find c and complete the equation.

The lines will intersect when they have the same y-coordinate. You can use this to form and solve an equation for x.

Then substitute this value of x to find y.

Answers

Q5
(a) The gradient of AB
$= \frac{-2-4}{6-(-2)} = -\frac{3}{4}.$

The equation is $y = -\frac{3}{4}x + c.$

$x = -2$ and $y = 4$ gives $c = \frac{5}{2}.$

$y = -\frac{3}{4}x + \frac{5}{2}$ or $3x + 4y - 10 = 0$

(b) The coordinates of D are
$\left(\frac{-2+6}{2}, \frac{4+(-2)}{2}\right) = (2, 1).$

The gradient of CD
$= \frac{5-1}{5-2} = \frac{4}{3}$

Product of gradients
$= -\frac{3}{4} \times \frac{4}{3} = -1,$ ∴ the lines are perpendicular.

Q6
(a) $(x-4)^2 + (y-(-2))^2 = (\sqrt{8})^2$
$(x-4)^2 + (y+2)^2 = 8$

(b) $y - x + 6 = 0$
$y = x - 6$
$(x-4)^2 + (x-6+2)^2 = 8$
$(x-4)^2 + (x-4)^2 = 8$
$x^2 - 8x + 16 + x^2 - 8x + 16 = 8$
$2x^2 - 16x + 24 = 0$
$x^2 - 8x + 12 = 0$
$(x-2)(x-6) = 0$
$x = 2$ or $x = 6$
Coordinates are (2, −4) and (6, 0).

Q7
$x^2 - 14x + y^2 - 10y + 49 = 0$
$(x-7)^2 - 49 + (y-5)^2 - 25 + 49 = 0$
$(x-7)^2 + (y-5)^2 = 25 = 5^2$
Centre is (7, 5).
Radius is 5.

How to solve these questions

First, find the gradient of the line AB.
This will allow you to write down the equation of the line.

Use the coordinates of A to find c and complete the equation of the line.

First find the coordinates of the midpoint.

Then you can calculate the gradient of CD.

You must show that the product of the gradients is -1.

Substitute the coordinates and the radius into the general equation of a circle.

First write the equation of the line in the form $y = \ldots$. Then substitute this into the equation of the circle and simplify.

Once you have a quadratic equation, this can be solved to give the x-coordinates.

Finally calculate the corresponding y-coordinates.

Use completing the square to write the equation in the form $(x-a)^2 + (y-b)^2 = r^2$
The centre (a, b) and the radius r can then be written down.

Answers

Q8

Centre has coordinates
$$\left(\frac{10+2}{2}, \frac{5+11}{2}\right) = (6, 8)$$

The distance between A and B is given by
$$\sqrt{(10-2)^2 + (11-5)^2} = \sqrt{64+36} = 10$$

Radius is 5.
Equation is $(x-6)^2 + (y-8)^2 = 5^2$.

Q9

(a) $x^2 + y^2 - 12x = 0$
$(x-6)^2 - 36 + y^2 = 0$
$(x-6)^2 + y^2 = 36 = 6^2$
Centre is (6, 0).
Radius is 6.

(b) $x^2 + (2x)^2 - 12x = 0$
$x^2 + 4x^2 - 12x = 0$
$5x^2 - 12x = 0$
$x(5x - 12) = 0$
$x = 0$ or $x = 2.4$
Intersects at **(0, 0)** and
(2.4, 4.8)

5 THE BINOMIAL EXPANSION

Q1

$(1 + 2x)^3$
$= 1 + 3 \times (2x) + 3 \times (2x)^2 + (2x)^3$
$= 1 + 6x + 12x^2 + 8x^3$

Q2

$(4 - x)^5$
$= 4^5 + 5 \times 4^4(-x) + 10 \times 4^3(-x)^2$
$+ 10 \times 4^2(-x)^3 + 5 \times 4(-x)^4 + (-x)^5$
$= 1024 - 1280x + 640x^2$
$- 160x^3 + 20x^4 - x^5$

Q3

$(5 - 3x)^4$
$= 5^4 + 4 \times 5^3(-3x) + 6 \times 5^2(-3x)^2$
$+ 4 \times 5(-3x)^3 + (-3x)^4$
$= 625 - 1500x + 1350x^2$
$- 540x^3 + 81x^4$

How to solve these questions

First find the coordinates of the centre which will be at the mid-point of AB.

Calculate the diameter of the circle and hence state the radius.

These values can now be substituted into the standard formula for the equation of a circle.

In this case it is only necessary to complete the square for the terms involving x. Once this has been done the equation will be in the standard form and you can state the coordinates of the centre and the radius.

In this case the first form of the equation is used, and $y = 2x$ is substituted into this. The equation is then solved to find two values for x.

Finally the corresponding y-coordinates are calculated.

Note that the coefficients 1, 3, 3, 1 can be obtained from Pascal's triangle.

Note that the coefficients 1, 5, 10, 10, 5, 1 can be obtained from Pascal's triangle.

Be careful with the negative signs.

Note that the coefficients 1, 4, 6, 4, 1 can be obtained from Pascal's triangle.

Answers

Q4

$\binom{5}{4} \times 2^1 \times (3x)^4$
$= \dfrac{5 \times 4 \times 3 \times 2 \times 1}{4 \times 3 \times 2 \times 1 \times 1} \times 2 \times 81x^4$
$= 810x^4$
The coefficient of x^4 is **810**.

Q5

$\left(x - \dfrac{1}{x}\right)^3 = x^3 + 3x^2\left(\dfrac{-1}{x}\right) + 3x$
$\times \left(\dfrac{-1}{x}\right)^2 + \left(\dfrac{-1}{x}\right)^3$
$= x^3 - 3x + \dfrac{3}{x} - \dfrac{1}{x^3}$

Q6

$(1 + x)^5 = 1 + 5x + 10x^2 + 10x^3$
$+ 5x^4 + x^5$

$(1 + y + y^2)^5$
$= 1 + 5(y + y^2) + 10(y + y^2)^2$
$+ 10(y + y^2)^3 + 5(y + y^2)^4$
$+ (y + y^2)^5$
$= 1 + 5y(1 + y) + 10y^2(1 + y)^2$
$+ 10y^3(1 + y)^3 + 5y^4(1 + y)^4$
$+ y^5(1 + y)^5$
$= \ldots + 10y^4 + \ldots + 10y^3 \times 3y$
$+ \ldots + 5y^4 \times 1 + \ldots$
$= \ldots + 45y^4 + \ldots$

The coefficient of y^4 is **45**.

Q7

$(1 - ax)^n$
$= 1 - nax + \dfrac{n(n-1)}{2}(ax)^2 - \ldots$

$na = 20 \Rightarrow a = \dfrac{20}{n}$

$\dfrac{n(n-1)}{2}a^2 = 160$

$\dfrac{n(n-1)}{2}\left(\dfrac{20}{n}\right)^2 = 160$

$200(n-1) = 160n$
$n = 5$
$a = \dfrac{20}{5} = 4$

How to solve these questions

You do not need to produce the whole expansion, only the x^4 term is needed.

State the coefficient clearly.

Use the formula for $(a + b)^n$ with $a = x$ and $b = -\dfrac{1}{x}$.

Be careful with the '–' signs as you simplify the expression.

Use the formula for the expansion of $(a + b)^n$, with $a = 1$ and $b = x$.

Replace each x in the expansion by $y + y^2$.

Note that $y, y^2, y^3, \ldots$ can be factored out of each bracket.

Only calculate the coefficients of the terms that will give y^4, but don't miss any out.

Generate the first 3 terms of the expansion in terms of a and n using
$$(1 + x)^n = 1 + nx + \frac{n(n-1)}{1 \times 2}x^2 + \ldots + \frac{n(n-1)\ldots(n-r+1)}{1 \times 2 \times \ldots \times r}x^r + \ldots$$

Equate the coefficients of x in both expressions to give one equation that relates a and n.

Repeat with coefficients of x^2 to form a second equation.

Combine the two equations to eliminate a and then solve for n.

Use the value of n to find a, by substituting it back into one of the earlier equations.

6 DIFFERENTIATION

Q1 $\dfrac{dy}{dx} = 4x^3 - 6x - 1$

Q2 $\dfrac{dy}{dx} = 3x^2 + 6$

Q3 $f'(x) = \dfrac{3}{2}x^{\frac{1}{2}} - x^{\frac{3}{2}}$

Q4 $f'(x) = 2x - \dfrac{16}{x^2}$

$2x - \dfrac{16}{x^2} > 0$

$x^3 > 8$

$x > 2$

Q5
(a) $y = x^3 - 3x^2 + 6x - 18$

$\dfrac{dy}{dx} = 3x^2 - 6x + 6$

(b) $15 = 3x^2 - 6x + 6$

$0 = 3(x-3)(x+1)$

$x = 3$ or $x = -1$

Q6 $\dfrac{dy}{dx} = 18x - 12x^2$

$18x - 12x^2 = 0$

$x = 0$ or $x = 1.5$

x	-1	0	1	1.5	2
$\dfrac{dy}{dx}$	-30	0	6	0	-12

Local minimum at (0, 0) and local maximum at (1.5, 6.75)

$18x - 12x^2 < 0$

$6x(3 - 2x) < 0$

$x < 0$ or $x > \dfrac{3}{2}$

How to solve these questions

To differentiate $f(x)$, you first need to write each term in the form x^n. In this case you should find
$f(x) = x\sqrt{x} + \dfrac{2\sqrt{x}}{x} = x^{\frac{3}{2}} + 2x^{-\frac{1}{2}}$ which you can then differentiate.

Differentiate $f(x)$ to find $f'(x)$.

For an increasing function you will have $f'(x) > 0$, so you need to form and then solve this inequality.

First you need to expand the brackets, and then you differentiate each term.

When the gradient is 15, $\dfrac{dy}{dx} = 15$, so you need to form and solve this equation.

You can solve the quadratic by factorising.

You first need to differentiate to find $\dfrac{dy}{dx}$.

To find the stationary points you need to form and solve the equation obtained when $\dfrac{dy}{dx} = 0$.

You should think about how the gradients change at each stationary point. It will help if you draw up a table like this.

To find the set of values for which a function is decreasing you first need to find $\dfrac{dy}{dx}$ and then form and solve the inequality $\dfrac{dy}{dx} < 0$.

Answers

Q7
$y = 4x - x^{\frac{1}{2}}$

$\dfrac{dy}{dx} = 4 - \dfrac{1}{2}x^{-\frac{1}{2}} = 4 - \dfrac{1}{2\sqrt{x}}$

When $x = 4$, $\dfrac{dy}{dx} = 4 - \dfrac{1}{4} = \dfrac{15}{4}$

For the tangent
$y = \dfrac{15}{4}x + c$

$x = 4, y = 14 \Rightarrow c = -1$

$y = \dfrac{15}{4}x - 1$

For the normal
$y = -\dfrac{4}{15}x - c$

$x = 4, y = 14 \Rightarrow c = \dfrac{226}{15}$

$y = -\dfrac{4}{15}x + \dfrac{226}{15}$

Q8
(a) Length $= 2x + 5$

Width $= \dfrac{20 - 2x}{2}$

$= 10 - x$

$A = (2x + 5)(10 - x)$

(b) $\dfrac{dA}{dx} = 15 - 4x$

$15 - 4x = 0 \Rightarrow x = 3.75$

x	3	3.75	4
$\dfrac{dA}{dx}$	3	0	-1

$A = (2 \times 3.75 + 5)(10 - 3.75)$

$= 78.1\ \text{m}^2$ (to 3 s.f.)

Q9
(a) Perimeter $= 2x + 2y + 2y + \pi x$

$= 2x + 4y + \pi x$

(b) Area $= 2x \times 2y + \dfrac{1}{2}\pi x^2$

$= 4xy + \dfrac{1}{2}\pi x^2$

(c) $100 = 2x + 4y + \pi x$

$y = \dfrac{100 - 2x - \pi x}{4}$

How to solve these questions

First differentiate and then calculate the value of $\dfrac{dy}{dx}$ when $x = 4$. This will give the gradient of the tangent.

Use the coordinates of the point on the curve to find the value of c.

The gradient of the normal is given by $-1 \Big/ \dfrac{dy}{dx}$.

Again, find the value of c, by using the coordinates of the point that you were given.

To find the area you need to find expressions for the length and the width.

You can then multiply these together.

You should first expand the brackets and find the derivative.

To find the value of x you must solve the equation $\dfrac{dy}{dx} = 0$.
You can use this table to show that gradient changes from positive to negative at $x = 3.75$, so that you have a local maximum.

You can substitute $x = 3.75$ into the formula for A to find the maximum value.

You find the perimeter by considering a semicircle and three of the sides of the rectangle.

You find the area by considering a semicircle and a rectangle.

You should use the equation for the perimeter to express y in terms of x.

Answers

$A = 4x\left(\dfrac{100 - 2x - \pi x}{4}\right) + \dfrac{1}{2}\pi x^2$

$= 100x - 2x^2 - \dfrac{1}{2}\pi x^2$

(d) $\dfrac{dA}{dx} = 100 - 4x - \pi x$

$0 = 100 - 4x - \pi x$

$x = \dfrac{100}{4+\pi} \Rightarrow y = 7.0\,\text{m}$

(e) $A = \dfrac{5000}{4+\pi}$

$= 700\,\text{m}^2$ (2 s.f.)

Maximum since
$\dfrac{d^2A}{dx^2} = -4 - \pi < 0$

Q10

(a) $x^2\sqrt{x} = x^2 \times x^{\frac{1}{2}} = x^{\frac{5}{2}}$

(b) $y = 2x^2 + x^{-4}$

$\dfrac{dy}{dx} = 2 \times \dfrac{5}{2}x^{\frac{3}{2}} - 4x^{-5} = 5x^{\frac{3}{2}} - \dfrac{4}{x^5}$

(c) When $x = 1$, $\dfrac{dy}{dx} = 5 - 4 = 1$

and $y = 2 + 1 = 3$.

Equation of tangent is

$y = x + c$.

$x = 1, y = 3 \Rightarrow c = 2$

$y = x + 2$

(d)

(i) $\dfrac{d^2y}{dx^2} = 5 \times \dfrac{3}{2}x^{\frac{1}{2}} + 20x^{-6}$

$= \dfrac{15}{2}\sqrt{x} + \dfrac{20}{x^6}$

(ii) As both $\sqrt{x}$ and x^6 are positive for every value of x that is greater than zero, then $\dfrac{d^2y}{dx^2}$ will also be positive, so C can only have a local minimum and never a local maximum.

Q11

(a) $\dfrac{dy}{dx} = 6x^2 - 6x - 12$

$6x^2 - 6x - 12 = 0$

$x^2 - x - 2 = 0$

$(x-2)(x+1) = 0$

$x = 2$ or $x = -1$

$(2, -27), (-1, 0)$

How to solve these questions

Then you can substitute for y in the expression for the area and simplify it to obtain the required result.
You could also collect the x^2-terms and write it as

$100x - (2 + \tfrac{1}{2}\pi)x^2$.

You should form and solve the equation $\dfrac{dA}{dx} = 0$.
It is often a good idea to leave answers in the exact form like this.

You can substitute the value of x into the formula for y to find the corresponding value for y.

Then substitute a value for x in the expression for A to obtain the actual area.

Remember that $\sqrt{x} = x^{\frac{1}{2}}$ and the rule for adding indices when multiplying.
Use the standard rule for differentiating x^n.

Calculate both the value of the derivative and y when $x = 1$. The derivative gives the gradient of the tangent. Use the values of x and y to find the constant c and then complete the equation.

Differentiate again to obtain $\dfrac{d^2y}{dx^2}$.

If there is to be no local maximum, the second derivative must always be greater than zero.

First differentiate and solve the equation $\dfrac{dy}{dx} = 0$.

Here the equation has been simplified and factorised, but you could use other methods to solve the equation.

Note that the y-coordinates have also been calculated at this stage.

Answers

(b) $\dfrac{d^2y}{dx^2} = 12x - 6$

If $x = 2$, $\dfrac{d^2y}{dx^2} = 24 - 6 = 18$

Local minimum at $(2, -27)$

If $x = -1$, $\dfrac{d^2y}{dx^2} = -12 - 6 = -18$

Local maximum at $(-1, 0)$

(c) $(x+1)^2(2x-7)$

$= (x^2 + 2x + 1)(2x - 7)$

$= 2x^3 + 4x^2 + 2x - 7x^2 - 14x - 7$

$= 2x^3 - 3x^2 - 12x - 7$

(d) $(x+1)^2(2x-7) = 0$

$x = -1$ or $x = \dfrac{7}{2}$

[Graph showing turning points with intersections at $(-1, 0)$, $(0, -7)$, $(2, -27)$, $(3.5, 0)$]

7 INTEGRATION

Q1 $\displaystyle\int_0^3 (2x^3 - x^2)\,dx = \left[\dfrac{x^4}{2} - \dfrac{x^3}{3}\right]_0^3$

$= \left(\dfrac{3^4}{2} - \dfrac{3^3}{3}\right) - 0$

$= 31.5$

Q2 $\displaystyle\int \dfrac{x^2 - x}{\sqrt{x}}\,dx = \int \left(x^{\frac{3}{2}} - x^{\frac{1}{2}}\right)dx$

$= \dfrac{2}{5}x^{\frac{5}{2}} - \dfrac{2}{3}x^{\frac{3}{2}} + c$

Q3 $x^2 = 12 - x$

$x^2 + x - 12 = 0$

$x = 3$ or $x = -4$

Area $= \displaystyle\int_0^3 x^2\,dx + \int_3^{12}(12 - x)\,dx$

$= \left[\dfrac{x^3}{3}\right]_0^3 + \left[12x - \dfrac{x^2}{2}\right]_3^{12}$

$= 9 + 40.5 = 49.5$

How to solve these questions

The second derivative has been found and used to determine the nature of the stationary points.

For a local maximum $\dfrac{d^2y}{dx^2} < 0$.

For a local minimum $\dfrac{d^2y}{dx^2} > 0$.

Multiply out the right hand side being sure to show enough working to convince the examiner.

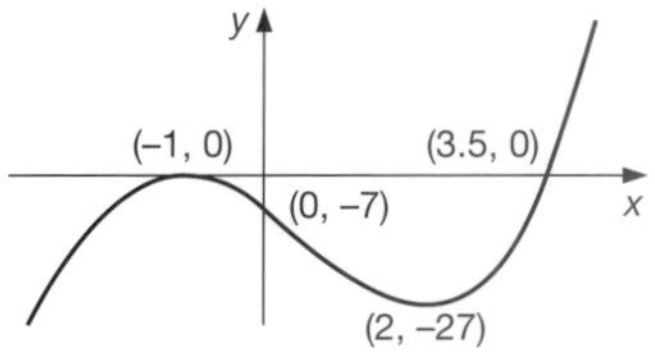

The graph should show the two turning points, the intersections with the axes at $(0, -7)$, $(-1, 0)$ and $\left(\dfrac{7}{2}, 0\right)$.

You should simplify the expression so that it consists of terms of the form x^n. Include a constant of integration.

You first need to find where the two curves intersect, by equating the two equations.

The graph shows the curve and the line.

You need to evaluate two integrals to find the area.

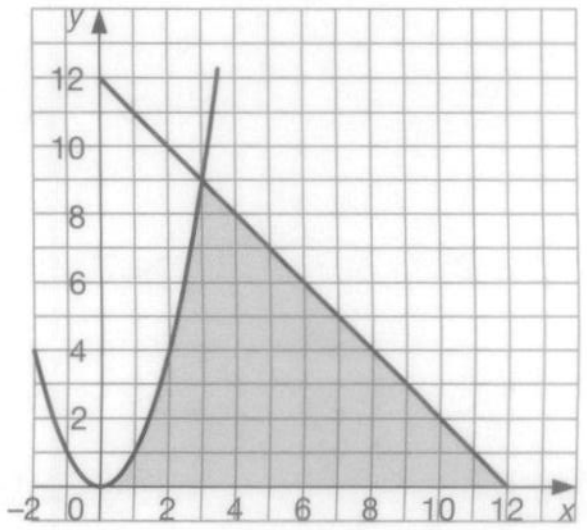

[Graph showing a curve and line with axes marked up to 12, with shaded area]

Answers (left)

Area $= \int_0^3 x^2 dx + \frac{1}{2} \times 9 \times 9$

$= 9 + 40.5 = 49.5$

Q4 $x^2 - 5x + 9 = 3$

$x^2 - 5x + 6 = 0$

$x = 2$ or $x = 3$

Area $= 3 \times 1 - \int_2^3 (x^2 - 5x + 9)\,dx$

$= 3 - \left[\frac{x^3}{3} - \frac{5x^2}{2} + 9x\right]_2^3$

$= 3 - \left(\frac{27}{2} - \frac{32}{3}\right)$

$= \frac{1}{6}$

Q5 Area $= \int_{-2}^1 (16 - x^4)\,dx - \frac{1}{2} \times 3 \times 15$

$= \left[16x - \frac{x^5}{5}\right]_{-2}^1 - 22.5$

$= 15.8 - (-25.6) - 22.5$

$= 18.9$

Q6 Area under the curve

$= \int_1^4 (5x^2 - x^3)\,dx - \frac{1}{2} \times 3(4 + 16)$

$= \left[\frac{5x^3}{3} - \frac{x^4}{4}\right]_1^4 - 30$

$= \frac{128}{3} - \frac{17}{12} - 30$

$= \frac{45}{4}$

How to solve these questions (left)

Alternatively, you can find the area under the curve and add the area of the triangle.

You must solve this equation to find the limits of integration. The region is shown in the diagram.

The area is found by subtracting the value of the integral from the area of the rectangle with vertices at the points (2, 3), (3, 3), (3, 0) and (2, 0).

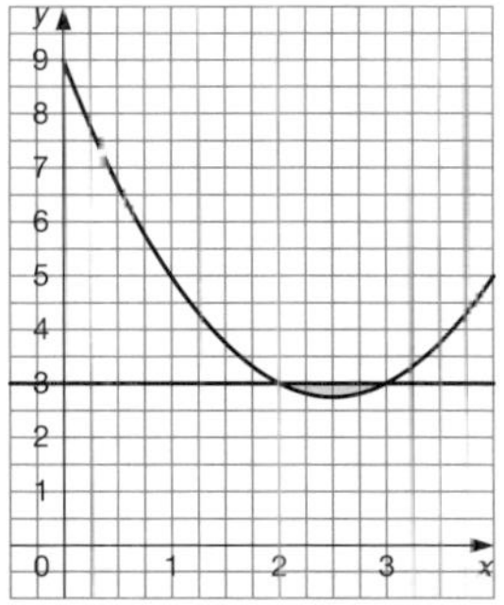

The diagram shows the area that must be found.

The area is given by the integral *minus* the area of the triangle between the line and the x-axis.

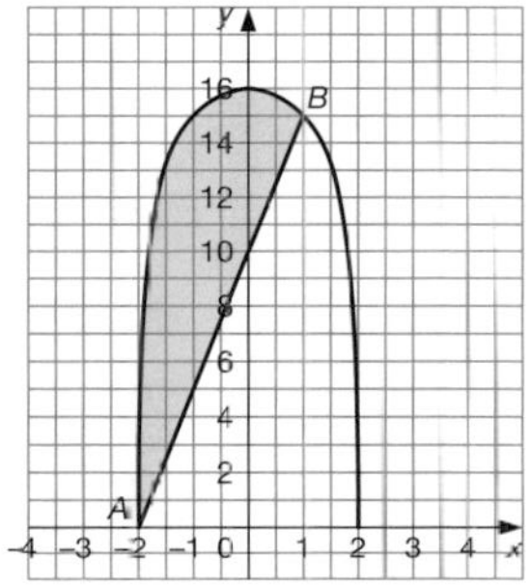

The diagram shows the area that must be found.

The area is given by the integral *minus* the area of the trapezium between the line and the x-axis.

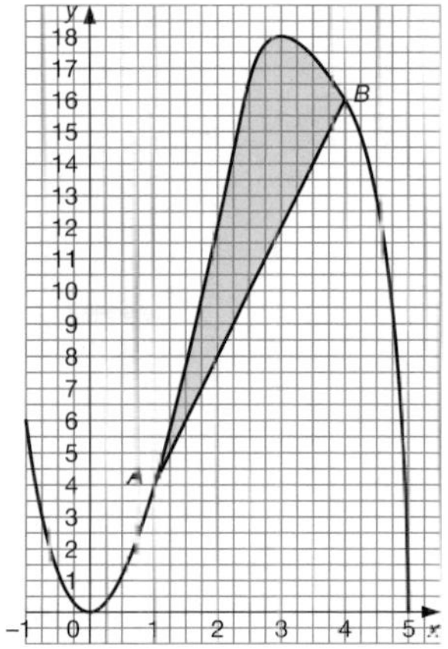

Answers (right)

Q7 $\int_0^2 (x^3 - 5x^2 + 6x)\,dx$

$= \left[\frac{x^4}{4} - \frac{5x^3}{3} + 3x^2\right]_0^2 = \frac{8}{3}$

$\int_2^3 (x^3 - 5x^2 + 6x)\,dx$

$= \left[\frac{x^4}{4} - \frac{5x^3}{3} + 3x^2\right]_2^3$

$= \frac{9}{4} - \frac{8}{3} = -\frac{5}{12}$

Area $= \frac{8}{3} + \frac{5}{12} = \frac{37}{12}$

Q8

(a) $\int_0^4 x(4 - x)\,dx$

$= \int_0^4 (4x - x^2)\,dx = \left[2x^2 - \frac{x^3}{3}\right]_0^4$

$= 2 \times 4^2 - \frac{4^3}{3} = \frac{32}{3}$

(b) (i) Area under curve

$= \int_0^k x(4 - x)\,dx$

$= \int_0^k (4x - x^2)\,dx$

$= \left[2x^2 - \frac{x^3}{3}\right]_0^k$

$= 2k^2 - \frac{k^3}{3}$

Area of triangle

$= \frac{1}{2} \times k \times k(4 - k)$

$= 2k^2 - \frac{k^3}{2}$

Area shaded

$= \left(2k^2 - \frac{k^3}{3}\right) - \left(2k^2 - \frac{k^3}{2}\right)$

$= \frac{k^3}{6}$

(ii) $\frac{k^3}{6} = \frac{1}{2} \times \frac{32}{3}$

$k = \sqrt[3]{32}$

$= 3.175$

How to solve these questions (right)

You should find the area of each of the two regions.

As the first region is above the axis you obtain a positive result from the integration.

As the second region is below the axis you will find a negative result, when you integrate. The area shaded below the x-axis is $\frac{5}{12}$.

Finally you should calculate the total area.

You need to expand the brackets before integrating.

Use this integral to calculate the total area under the curve between x = 0 and x = k.

You should then express the area of the triangle in terms of k.

Finally you can find the difference to give the shaded area.

From part (a), you know that the total area under the curve from x = 0 to x = 4 is $\frac{32}{3}$. You also know that the area shaded is $\frac{k^3}{6}$. Form and solve this equation to find k.

Answers

Q9

(a) $\int_{-1}^{2}(x^2 - 2x + 2)\,dx$

$= \left[\dfrac{x^3}{3} - x^2 + 2x\right]_{-1}^{2}$

$= \left(\dfrac{2^3}{3} - 2^2 + 2\times 2\right)$

$\quad - \left(\dfrac{(-1)^3}{3} - (-1)^2 + 2\times(-1)\right)$

$= \dfrac{8}{3} - \left(\dfrac{-10}{3}\right)$

$= \dfrac{18}{3} = 6$

(b) Area of trapezium

$= \dfrac{1}{2}(2+5)\times 3$

$= \dfrac{21}{2}$

$= 10.5$

Area of R

$= 10.5 - 6$

$= 4.5$

8 THE TRAPEZIUM RULE

Q1

x	0	0.25	0.5	0.75	1
2^x	1	1.1892	1.4142	1.6818	2

$\int_0^1 2^x dx \approx \dfrac{1}{2}(1 + 2\times 1.1892 + 2\times 1.6818 + 2)\times 0.25$

$1.4142 + 2\times 1.6818 + 2)\times 0.25$

$= 1.4$

Q2

x	1	1.2	1.4	1.6	1.8	2
$\dfrac{2}{x}$	2	1.667	1.429	1.25	1.111	1

$\int_1^2 \dfrac{2}{x} dx \approx \dfrac{1}{2}(2 + 2\times 1.667 + 2\times 1.429 + 2\times 1.25 + 2\times 1.111 + 1)\times 0.2$

$= 1.4$

How to solve these questions

Use the standard integration techniques.

Use the formula for the area of a trapezium $A = \dfrac{1}{2}(a+b)h$.

Subtract the integral from the area of the trapezium to find the area of the region R.

Complete the table given in the question to give a list of the ordinates.

Use the trapezium rule with these ordinates.

Complete the table to give you the required ordinates.

Apply the trapezium rule to these ordinates.

Answers

Q3

$h = \dfrac{2-1}{5} = 0.2$

x	1	1.2	1.4	1.6	1.8	2
$x + \dfrac{1}{x}$	2	2.033	2.114	2.225	2.356	2.5

$\int_1^2 x + \dfrac{1}{x}\,dx \approx \dfrac{1}{2}(2 + 2\times 2.033 + 2\times 2.114 + 2\times 2.225 + 2\times 2.356 + 2.5)\times 0.2$

$= 2.2$

Q4

(a)

$h = \dfrac{3-1}{2} = 1$

x	1	2	3
$\dfrac{1}{\sqrt{x}}$	1	0.7071	0.5774

$\int_1^3 \dfrac{1}{\sqrt{x}}\,dx \approx \dfrac{1}{2}(1 + 2\times 0.7071 + 0.5774)\times 1$

$= 1.50$

(b)

$h = \dfrac{3-1}{4} = 0.5$

x	1	1.5	2	2.5	3
$\dfrac{1}{\sqrt{x}}$	1	0.8165	0.7071	0.6325	0.5774

$\int_1^3 \dfrac{1}{\sqrt{x}}\,dx \approx \dfrac{1}{2}(1 + 2\times 0.8165 + 2\times 0.7071 + 2\times 0.7071 + 2\times 0.6325 + 0.5774)\times 0.5$

$= 1.47$

(c)

$\int_1^3 \dfrac{1}{\sqrt{x}}\,dx = \int_1^3 x^{\frac{1}{2}}\ dx = \left[2x^{\frac{1}{2}}\right]_1^3 = 2\sqrt{3} - 2$

$= 1.46$

(d) The second estimate is better as it uses narrower trapeziums and so reduces the differences between the curve and the trapeziums.

How to solve these questions

First calculate the width of each trapezium.

Calculate the ordinates, based on this trapezium width.

Apply the trapezium rule to these ordinates.

Calculate the value of h.

Evaluate the ordinates.

Apply the trapezium rule.

Repeat with the larger number of ordinates.

Evaluate the definite integral using

$$\int x^n dx = \dfrac{x^{n+1}}{n+1} + c$$

and give your answer to two decimal places.

Explain why the second estimate is better.

9 FACTOR AND REMAINDER THEOREMS

Q1

```
          x² -  x -  6
x - 4 )x³ - 5x² - 2x + 24
       x³ - 4x²
       -x² - 2x
       -x² + 4x
            -6x + 24
            -6x + 24
                    0
```

Multiply $(x-4)$ by x^2, then $-x$ and then by -6.

Q2

(a)

```
          x² + 2x - 8
x + 1 )x³ + 3x² - 6x - 8
       x³ +  x²
            2x² - 6x
            2x² + 2x
                 -8x - 8
                 -8x - 8
                       0
```

Use long division, first multiplying $(x + 1)$ by x^2, then $-x$ and then -8.

(b) $x^3 + 3x^2 - 6x - 8 = 0$
$(x + 1)(x^2 + 2x - 8) = 0$
$(x + 1)(x - 2)(x + 4) = 0$
$x = -1$ or $x = 2$ or $x = -4$

The equation can be written as the product of a linear and a quadratic term. The quadratic term can be factorised and the solutions can then be stated.

Q3

(a) $(-3)^3 - 3 \times (-3)^2 - 25 \times (-3) - 21$
$= -27 - 27 + 75 - 21$
$= 0$
$\therefore (x + 3)$ is a factor

Substitute $x = -3$ and check that this gives a value of zero.

(b)

```
           x² - 6x - 7
x + 3 )x³ - 3x² - 25x - 21
       x³ + 3x²
           -6x² - 25x
           -6x² - 18x
                  -7x - 21
                  -7x - 21
                        0
```

$x^3 - 3x^2 - 25x - 21$
$= (x + 3)(x^2 - 6x - 7)$
$= (x + 3)(x + 1)(x - 7)$

Once you have identified one factor you can use long division to find a quadratic factor.

(c) $x = -3$ or $x = -1$ or $x = 7$

The quadratic factor can then in turn be factorised to give the three linear factors.

Q4

(a) $1 - 4 - k + 10 = 0$
$7 - k = 0$
$k = 7$

Given that $(x - 1)$ is a factor, then $f(1) = 0$. This can be used to form an equation to find k.

(b)

```
           x² - 3x - 10
x - 1 )x³ - 4x² - 7x + 10
       x³ -  x²
           -3x² - 7x
           -3x² + 3x
                -10x + 10
                -10x + 10
                        0
```

A long division can be carried out to express f(x) as the product of a linear and a quadratic term.

(c) $x^3 - 4x^2 - 7x + 10$
$= (x - 1)(x^2 - 3x - 10)$
$= (x - 1)(x + 2)(x - 5)$

The quadratic term can be factorised to give the three linear factors.

(d) $f(2) = 1 \times 4 \times (-3) = -12$
Remainder is -12.

The remainder theorem states that the remainder when f(x) is divided by $(x - 2)$ is equal to f(2).

Q5

(a) $64 - 16a + 92 + 36 = 0$
$192 - 16a = 0$
$a = \dfrac{192}{16} = 12$

As $(x - 4)$ is a factor, $f(4) = 0$. This allows an equation to be formed and solved to find a.

(b) $f(2) = 8 - 12 \times 4 + 23 \times 2 + 36 = 42$
Remainder is 42.

The remainder will be given by f(2).

(c) $f(9) = 729 - 972 + 207 + 36 = 0$
So $(x - 9)$ is a factor of f(x).

To show that $(x - 9)$ is a factor, you need to show that f(9) = 0.

Q6

(a) $f(-1) = -1 - 9 - 14 + 24 = 0$
so $(x + 1)$ is a factor.

Show that $f(-1) = 0$, then by the factor theorem $(x + 1)$ is a factor.

(b)

```
            x² - 10x + 24
x + 1 )x³ - 9x² + 14x + 24
       x³ +  x²
          -10x² + 14x
          -10x² - 10x
                  24x + 24
                  24x + 24
                        0
```

Long division should then be carried out to find a quadratic factor.

(c) $x^3 - 9x^2 + 14x + 24$
$= (x + 1)(x^2 - 10x + 24)$
$= (x + 1)(x - 4)(x - 6)$
$x = -1$ or $x = 4$ or $x = 6$

The quadratic term can then be factorised to give the three linear factors.

The solutions can be stated once the linear factors have been identified.

Answers

Q7

$1 - p - q + 14 = 0$
$p + q = 15$ (1)

$8 - 4p - 2q + 14 = -20$
$4p + 2q = 42$
$2p + q = 21$ (2)

            (2) – (1)
$p = 6$
$6 + q = 15$
$q = 9$

Q8

(a)
$8 + 4a + 2b - 6 = -8 + 4a - 2b - 6$
$8 + 2b = -8 - 2b$
$4b = -16$
$b = -4$

(b)
$1 + a - 4 - 6 = 0$
$a - 9 = 0$
$a = 9$

(c)

$$\begin{array}{r} x^2 + 10x + 6 \\ x-1 \overline{)\, x^3 + 9x^2 - 4x - 6} \\ \underline{x^3 - x^2} \\ 10x^2 - 4x \\ \underline{10x^2 - 10x} \\ 6x - 6 \\ \underline{6x - 6} \\ 0 \end{array}$$

$x^3 + 9x^2 - 4x - 6$
$= (x-1)(x^2 + 10x + 6)$

(d)
$10^2 - 4 \times 1 \times 6 = 76$
As this is positive the quadratic will have two factors and there will be three real solutions to the equation $f(x) = 0$.

How to solve these questions

A first equation can be formed using $f(1) = 0$ since $(x - 1)$ is a factor.

A second equation can be formed using $f(2) = -20$ since the remainder is -20 when $f(x)$ is divided by $(x - 2)$.

The pair of simultaneous equations can now be solved.

To find b form an equation using $f(2) = f(-2)$ since the two remainders are equal.

To find a form an equation using $f(1) = 0$.

Use long division to find the quadratic term.

The number of solutions of $f(x) = 0$ depends on the number of solutions of $x^2 + 10x + 6 = 0$. This can be determined using the discriminant of this quadratic.

10 EXPONENTIALS AND LOGARITHMS

Q1

(a) $\log_{10}1000000 = \log_{10}10^6 = 6$

(b) $\log_3 81 = \log_3 3^4 = 4$

(c) $\log_5 0.2 = \log_5 5^{-1} = -1$

(d) $\log_4 2 = \log_4 4^{\frac{1}{2}} = \frac{1}{2}$

Use the result $\log_a a^n = n$.

Note that $0.2 = \frac{1}{5} = 5^{-1}$.

Note that $2 = \sqrt{4} = 4^{\frac{1}{2}}$.

Answers

Q2

(a)
$\log_n(a^3) = 3\log_n a = 3 \times 4 = 12$

(b)
$\log_n(\sqrt{a}) = \frac{1}{2}\log_n a = \frac{1}{2} \times 4 = 2$

(c)
$\log_n\left(\frac{1}{a^2}\right) = -2\log_n a = -2 \times 4 = -8$

Q3
$2 = 1.4^x$
$\log 2 = \log(1.4^x)$
$\log 2 = x\log 1.4$
$x = \dfrac{\log 2}{\log 1.4} = 2.06$ (to 3 s.f.)

Q4

(a) $\log_n 32 = 5$
$n^5 = 32$
$n = 2$

(b) $\log_5 n = -2$
$n = 5^{-2} = \frac{1}{25} = 0.04$

(c) $\log_3 729 = n$
$729 = 3^n$
$\log 729 = n\log 3$
$n = \dfrac{\log 729}{\log 3} = 6$

Q5

(a) $\log_4 16 = 2$
$\log_7 49 = 2$
$\therefore$ **True**

(b) $\frac{1}{3}\log_2 8 = \frac{1}{3} \times 3 \times 1 = 1$
$\frac{1}{3}\log_4 256 = \frac{1}{3} \times 4 = \frac{4}{3}$
$\therefore$ **False**

(c) $\log_3 9 = 2$
$\log_{10}1000 \times \log_4 \sqrt[3]{16}$
$= \log_{10}1000 \times \frac{1}{3}\log_4 16$
$= 3 \times \frac{1}{3} \times 2$
$= 2$
$\therefore$ **True**

How to solve these questions

Use the rule $n\log a = \log(a^n)$.

Note that $\sqrt{a} = a^{\frac{1}{2}}$.

Note that $\frac{1}{a^2} = a^{-2}$.

Take logs of both sides of the equation. Use the rule $n\log a = \log(a^n)$.

Then solve for x.

In each case use the result $\log_a a^b = b$ to form an equation that can be used to find n.

Note that $\log_n n^2 = 2$.

Note that $\log_2 8 = 3$.

Note that $\log_4 256 = 4$.

Note that $\sqrt[3]{16} = 16^{\frac{1}{3}}$.

Q6

(a) $\log_2 x = \log_5 125 + \log_5 0.008$

$\log_2 x = 3 + (-3)$

$\log_2 x = 0$

$x = 1$

(b) $\log_n x + \log_n 15 = \log_n 900$

$\log_n 15x = \log_n 900$

$15x = 900$

$x = \dfrac{900}{15} = 60$

(c) $2\log_n x - \log_n x = \log_n (2 - x)$

$\log_n x = \log_n (2 - x)$

$x = 2 - x$

$2x = 2$

$x = 1$

(d) $\log_n x^3 + \log_n x^2 = 10\log_n 5$

$3\log_n x + 2\log_n x = 10\log_n 5$

$5\log_n x = 10\log_n 5$

$\log_n x = 2\log_n 5$

$\log_n x = \log_n 25$

$x = 25$

Q7

(a) $T = 80 \times 0.95^0 + 20 = 80 + 20$

$= 100$

(b) $30 = 80 \times 0.95^t + 20$

$10 = 80 \times 0.95^t$

$0.125 = 0.95^t$

$\log 0.125 = t\log 0.95$

$t = \dfrac{\log 0.125}{\log 0.95} = 40.5$ (to 3 s.f.)

(c) *T will get closer and closer to 20.*

Q8

$2^t + 4 \times 2^{-t} = 5$

$2^{2t} + 4 = 5 \times 2^t$

$2^{2t} - 5 \times 2^t + 4 = 0$

$(2^t - 1)(2^t - 4) = 0$

$2^t = 1$ or $2^t = 4$

$t = 0$ or $t = 2$

How to solve these questions

Note $125 = 5^3$ and $0.008 = 5^{-3}$.

Using the rule $\log_n a + \log_n b = \log_n (ab)$.

Using $n\log a = \log a^n$.

Note that $0.95^0 = 1$.

Substitute $T = 30$.

When the equation has been simplified it is possible to take logs of both sides and then solve for t.

Multiply every term by 2^t.

Then simplify and factorise.
Two values for t can then be obtained.

---

11 TRANSFORMATIONS

Q1

Q2

(a) f(−x) is a reflection in the y-axis.

(b) −f(x) is a reflection in the x-axis.

Q3

(a) $y = -f(x)$ — Reflection in the x-axis.

(b) $y = f(2x)$ — Horizontal stretch factor $\dfrac{1}{2}$ parallel to the x-axis.

(c) $y = f(x - 1)$ — Horizontal translation of 1 unit.

(d) $y = f(x) + 1$ — Vertical translation of 1 unit.

Answers

How to solve these questions

Q4

(a)

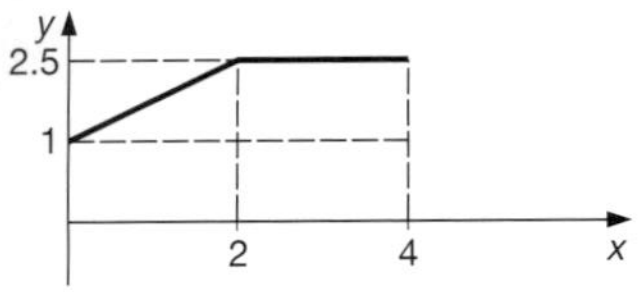

The graph has been moved one unit upwards.

(b)

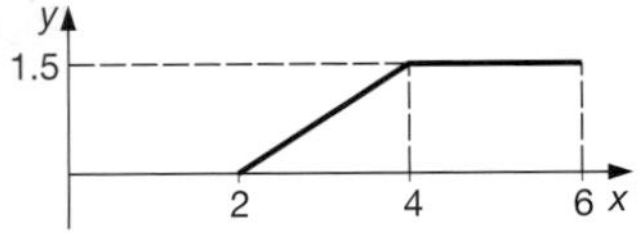

The graph has been moved two units to the right.

(c)

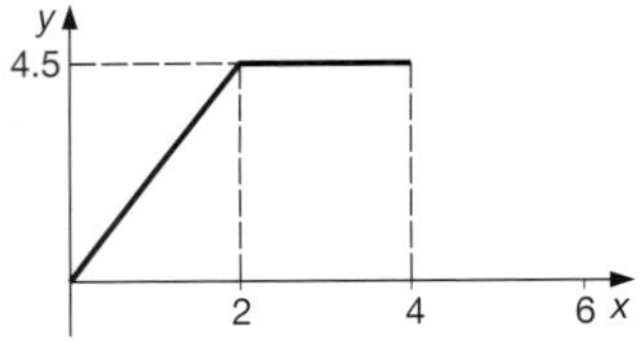

The graph has been stretched vertically by a factor of 3.

(d)

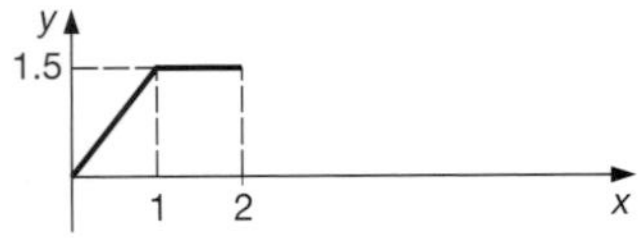

The graph has been stretched horizontally by a factor of $\frac{1}{2}$.

Answers

12 Kinematics on a straight line

Q1

(a) $0^2 = 6^2 + 2 \times (-9.8)s$

$s = \dfrac{36}{2 \times 9.8} = 1.84\,\text{m}$

Maximum height $= 5 + 1.84$

$= \textbf{6.84 m}$

(b) $-5 = 6t + \dfrac{1}{2}(-9.8)t^2$

$4.9t^2 - 6t - 5 = 0$

$t = 1.79$ or $t = -0.57$

Time in air $= \textbf{1.79 s}$

(c) $v^2 = 6^2 + 2 \times (-9.8) \times (-5)$

$v = \sqrt{134} = \textbf{11.58 m s}^{-1}$

Q2

(a) $v = 10 + 3 \times 6$

$= \textbf{28 m s}^{-1}$

(b) For OA:

$10^2 = 0^2 + 2 \times 4 \times s$

$s = 12.5\,\text{m}$

For AB:

$s = 10 \times 6 + \dfrac{1}{2} \times 3 \times 6^2$

$= 114\,\text{m}$

Total distance $= 12.5 + 114$

$= \textbf{126.5 m}$

Q3

(a)

(b) Distance $= \dfrac{1}{2}(90 + 120) \times 36$

$= \textbf{3780 m}$

How to solve these questions

Q1

First note that $a = -9.8$ and $v = 0$, at the maximum height of the ball.

Then use $v^2 = u^2 + 2as$.

You must add on the height from which the ball was thrown.

Note that when the ball hits the ground, $s = -5$ and use the equation $s = ut + \dfrac{1}{2}at^2$.

You must take the positive solution in this case.

Use $s = -5$ and the equation $v^2 = u^2 + 2as$.

Use $v = u + at$ with $u = 10$, $a = 3$ and $t = 6$ to find the speed.

Calculate the distance for each part.

You should use $v^2 = u^2 + 2as$.

Here you should use $s = ut + \dfrac{1}{2}at^2$.

Add the two distances to find OB.

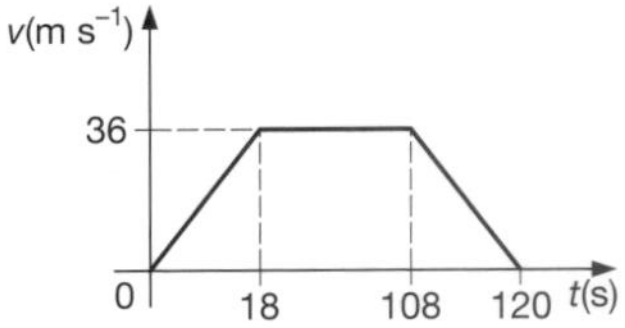

Your graph should show each stage of the journey and you should mark in the key times and the velocity as shown here.

You can find the distance by using the graph and finding the area.

Answers

(c) It travels at constant speed for most of its journey.

(d) $3780 = \frac{1}{2} \times 150 \times V$

$V = 50.4 \text{ m s}^{-1}$

Q4

(a) The train slows down, then travels at a constant speed before slowing down and coming to rest.

(b) 1st stage:

distance $= \frac{1}{2}(3 + 1.2) \times 5$

$= 10.5 \text{ m}$

2nd stage:

distance $= 1.2 \times 5$

$= 6 \text{ m}$

3rd stage:

distance $= \frac{1}{2} \times 6 \times 1.2$

$= 3.6 \text{ m}$

Total distance $= 10.5 + 6 + 3.6$

$= 20.1 \text{ m}$

(c) At $t = 12$, $v = 0.8 \text{ m s}^{-1}$

From $t = 10$ to 12:

$s = \frac{1}{2}(1.2 + 0.8) \times 2 = 2 \text{ m}$

Total distance $= 10.5 + 6 + 2$

$= 18.5 \text{ m}$

Q5

(a)

How to solve these questions

This is the most obvious difference.

The graph is in the shape of a triangle, which has height V and base 150.

You should identify the key features of the journey.

You should calculate the distance for each stage of the journey.

For the first stage you need to find the area of a trapezium, for the second stage the area of a rectangle and for the third stage the area of a triangle.

Then you can find the total distance.

First, calculate the speed when $t = 12$.

Then you can use $s = \frac{1}{2}(u + v)t$.

Again you will need to include the distance travelled in the first two stages.

Your graph should have three stages.

Make sure you label all the key values on each axis.

Answers

(b) $20 = \frac{1}{2} \times 2 \times 4 + t$

$+ \frac{1}{2} \times 1 \times 4$

$20 = 6 + 4t$

$t = 3.5 \text{ s}$

(c) Total time $= 2 + 3.5 + 1 = \textbf{6.5 s}$

Average velocity $= \frac{20}{6.5}$

$= 3.08 \text{ m s}^{-1}$

Q6

(a) Let V be the speed and T the time at B.

$V = 20 + 2T$

$AB = 20T + \frac{1}{2} \times 2T^2$

$= 20T + T^2$

$BC = 10V$

$= 10(20 + 2T)$

$= 200 + 20T$

$425 = 20T + T^2 + 200 - 20T$

$0 = T^2 + 40T - 225$

$T = 5$ or $T = -45$

Time to travel from A to B is 5 s.

(b) $V = 20 + 2 \times 5$

$= 30 \text{ m s}^{-1}$

13 KINEMATICS AND VECTORS

Q1 $i - 3j = 4i + 2j + a \times 10$

$10a = -3i - 5j$

$a = (-0.3i - 0.5j) \text{ m s}^{-2}$

Q2

(a) $r = (5i - 5j)t + \frac{1}{2}(-i + 2j)t^2$

$= (5t - \frac{1}{2}t^2)i + (-5t + t^2)j$

$v = (5i - 5j) + (-i + 2j)t$

$= (5 - t)i + (2t - 5)j$

How to solve these questions

You can form this equation by noting that the total area under the graph must be 20.

Don't forget that you need the total time.

Use:

$$\text{average velocity} = \frac{\text{total displacement}}{\text{total time}}$$

You should use $v = u + at$ here.

You should use $s = ut + \frac{1}{2}at^2$ here.

Find the distance BC in terms of V and then convert this so that it is in terms of T.

You can then use the fact that the total distance is 425 m to form and solve an equation.

You can use your earlier expression for V with $T = 5$.

You need to use the formula $v = u + at$ with $v = i - 3j$, $u = 4i + 2j$ and $t = 10$. Then solve to find a.

You should use the formula $r = ut + \frac{1}{2}at^2$ to find the position.

You should use the formula $v = u + at$ to find the velocity.

Answers

(b) When $5 - t = 0$
$t = 5$ s

How to solve these questions

When it is heading north the i-component of the velocity will be zero. You should use this to form and solve an equation.

Q3

(a) $r_A = (4i + 3j)t + 4j$
$= 4ti + (3t + 4)j$
$r_B = (2i - j)t + 4i + 12j$
$= (2t + 4)i + (12 - t)j$

How to solve these questions

The positions can be found by multiplying the velocity by t and adding the initial position.

(b) $4t = 2t + 4$
$t = 2$
$3t + 4 = 12 - t$
$t = 2$
∴ they collide when $t = 2$
The position at collision is $8i + 10j$.

You need both components to be equal at the time of the collision. This gives two equations that will have the same solution if a collision takes place.

(c) When $t = 1$:
$r_A = 4i + 7j$
$r_B = 6i + 11j$
$r_B - r_A = 2i + 4j$
Distance $= \sqrt{2^2 + 4^2}$
$= 4.47$ m (3 s.f.)

Then you can substitute this time into either position vector, to find where the boats collide.

You should first find the positions of both boats at this time.

Find the displacement of B relative to A.

Use Pythagoras' theorem to calculate the distance.

Q4

(a)

Your triangle should show the current, which has a velocity of 0.8 m s⁻¹, the resultant velocity, which is at 30° to the current, and the velocity of the swimmer which is 1.2 m s⁻¹, and that completes the triangle.

(b) $\dfrac{\sin \beta}{0.8} = \dfrac{\sin 30°}{1.2}$
$\beta = 19.47°$
$\alpha = 130.53°$
$v = 1.82$ m s⁻¹ (3 s.f.)

You need to find the length of the longest side which represents the resultant velocity. First find the angle β, using the sine rule.

Then you can calculate α.

Finally you can use the sine rule again to find the resultant velocity.

Answers

Q5

(a) $v = 3i$

How to solve these questions

As the boat is travelling due east, the velocity will be a multiple of i and as the speed is 3 m s⁻¹, the velocity is 3i.

(b) $r = (3i)t + \dfrac{1}{2}(-0.2i - 0.3j)t^2$
$= (3t - 0.1t^2)i - 0.15t^2j$

The position vector can be found using the formula $r = ut + \dfrac{1}{2}at^2$. The result has been simplified ready to start part (c).

(c) $3t - 0.1t^2 = 0$
$t(3 - 0.1t) = 0$
$t = 0$ or $t = \dfrac{3}{0.1} = 30$ s
$t = 30 \Rightarrow r = -135j$
Due south of origin.

When the boat is due south of the origin, the i component must be zero. This gives an equation that can be solved to find t. The position vector can be found to check that the boat is due south at this time.

(d) $v = 3i + (-0.2i - 0.3j)t$
$= (3 - 0.2t)i - 0.3tj$
$3 - 0.2t = 0.3t$
$t = \dfrac{3}{0.5} = 6$ s
$v = 1.8i - 1.8j$
Travelling south east.
$r = (3 \times 6 - 0.1 \times 6^2)i$
$\quad - 0.15 \times 6^2j$
$= 14.4i - 5.4j$
$r = \sqrt{14.4^2 + 5.4^2} = 15.4$ m

For this part of the question you need to use the velocity, so the first step is to write down the velocity. To travel south east the i- and j-components must have the same magnitude, with the i-component being positive and the j-component being negative. This gives an equation that can be solved to find t. This can then be used to check that the velocity is south east. Finally calculate the position vector at this time and then the magnitude of this vector, which gives the distance from the origin.

Q6

(a) $r = 20ti$
$s = (10i + 10j)t + 300i$
$= (10t + 300)i + 10tj$

The position vectors are given by the product of the velocity and t plus the initial positions.

(b) $\overrightarrow{AB} = s - r$
$= (300 - 10t)i + 10tj$

This is found by subtracting the position of A from the position of B.

When on this bearing both components of $\overrightarrow{AB}$ will be equal.

(c) $300 - 10t = 10t$
$t = 15$ s

(d) $(300 - 10t)^2 + (10t)^2 = 300^2$
$200t^2 - 6000t = 0$
$t = 0$ or 30
So they are 300 m apart again when $t = 30$ s.

Use Pythagoras' theorem to find the distance between A and B.

Then form and solve this quadratic equation.

How to solve these questions

14 NEWTON'S LAWS AND CONNECTED PARTICLES

Q1

(a) $W = 15 \times 9.8 = 147$ N

The weight is simply given by mg.

(b) $80 \sin 30° + R = 147$

$R = 147 - 80 \sin 30° = 107$ N

Resolve vertically to find the normal reaction. Note that it is not equal to the weight because of the vertical component of the tension.

(c) $F = 0.4 \times 107 = 42.8$ N

As the sledge is moving use $F = \mu R$ to find the friction force.

(d) $80 \cos 30° - 42.8 = 15a$

$a = \dfrac{80 \cos 30° - 42.8}{15} = 1.77\,\text{ms}^{-2}$

Resolve horizontally and apply Newton's second law to find the acceleration.

(e) $v = 0 + a \times 3 = 5.30\,\text{m s}^{-1}$

The velocity can be found using $v = u + at$. Use the exact value of a to ensure that your final result is correct.

Q2 *See diagram opposite.*

Show the forces clearly.

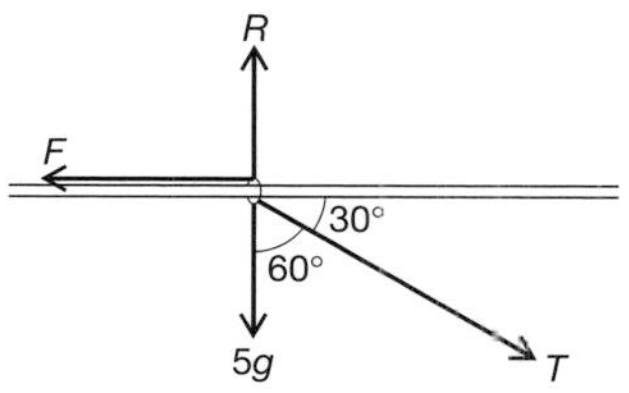

Resolving forces vertically:

$R = 5g + T\cos60°$ (1)

Resolving forces horizontally:

$F = T\cos30°$ (2)

Remember that the friction opposes the direction of the motion of the ring.

Using the law of limiting friction:

$F = \dfrac{1}{2} R$ (3)

Since the ring is about to slip, use limiting friction $F = \mu R$.

(3) and (2) $\Rightarrow T\cos30° = \dfrac{1}{2} R$

$R = 2T\cos30°$

From (1):

$5g + T\cos60° = 2T\cos30°$

$T = \dfrac{5g}{2\cos30° - \cos60°}$

$T = 39.8$ N

How to solve these questions

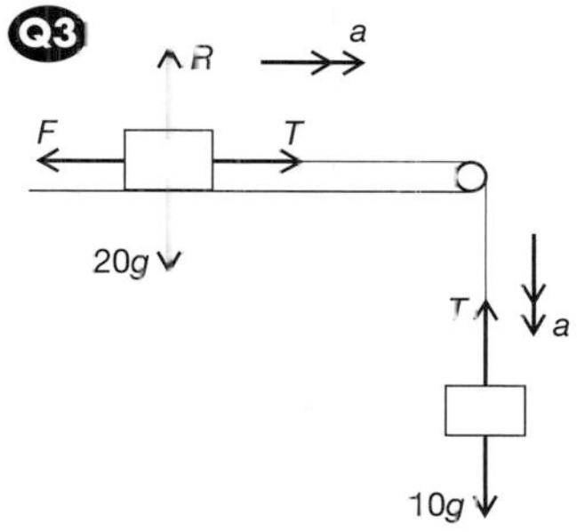

First, draw a diagram and show all the forces on the objects.

Q3

Applying Newton's second law to the 10-kg mass:

$98 - T = 10 \times 0.5$

$T = 93$ N

Apply Newton's second law to each object in the system separately.

Applying Newton's second law to the 20-kg mass:

Horizontally:

$93 - F = 20 \times 0.5$

$F = 83$ N

Vertically:

$R = 20 \times 9.8 = 196$ N

Using $F = \mu R$

$83 = \mu \times 196$

$\mu = \dfrac{83}{196} = 0.423$ (to 3 s.f.)

Q4

(a) $T \sin 48° = 60 \sin 50°$

$T = \dfrac{60 \sin 50°}{\sin 48°} = 61.8$ N

First resolve horizontally to find T.

(b) $60 \cos 50° + T \cos 48° = mg$

$m = \dfrac{60 \cos 50° + T \cos 48°}{9.8}$

$= 8.16\,\text{kg}$

To find m resolve vertically. Use the exact value of T from part (a), to avoid introducing any error into your final answer.

Use $g = 9.8\ \text{m s}^{-2}$ here.

Answers

Q5

(a) For the 3-kg particle
$$3g - T = 3a \quad (1)$$
For the 2-kg particle
$$T - 2g = 2a \quad (2)$$
From equation (2)
$$a = \frac{T - 2g}{2}$$

Substituting into equation (1) gives
$$3g - T = 3\left(\frac{T - 2g}{2}\right)$$
$$6g - 2T = 3T - 6g$$
$$5T = 12g$$
$$T = \frac{12g}{5} = 23.52 \text{ N}$$

(b) $a = \dfrac{T - 2g}{2} = \dfrac{23.52 - 2 \times 9.8}{2}$
$$= 1.96 \text{ m s}^{-2}$$

15 CONSERVATION OF MOMENTUM

Q1

(a) $5 \times 6 + 3 \times 4 = 5v + 3v$
$$42 = 8v$$
$$v = \frac{42}{8} = 5.25 \text{ m s}^{-1}$$

(b) $5 \times 6 + 3 \times (-4) = 5v + 3v$
$$18 = 8v$$
$$v = \frac{18}{8} = 2.25 \text{ m s}^{-1}$$

Q2 Total momentum before collision = 0

Total momentum after collision $= 48 \times 0.5 + 1.2v$
$$= 24 + 1.2v$$
$$0 = 24 + 1.2v$$
$$v = -20 \text{ m s}^{-1}$$

How to solve these questions

(a) It is important that you form an equation of motion for each of the particles.

The acceleration, a, can then be eliminated to find the tension as required.

You can find a first and then use this to find T if you find that easier.

(b) The value for T can then be substituted into one of the equations to find a.

In the first case use conservation of momentum with both of the velocities taken to be positive. Note that v is the velocity after the collision. As this is positive it is the same as the speed.

In this case the velocity of particle B is taken to be negative.

As both are at rest the total momentum is zero before the collision.

The negative sign indicates that the skateboard moves in the opposite direction to the child.

Answers

Q3

(a) Momentum before collision
$$= 2x + 0.1 \times (-3)$$
$$= 2x - 0.3$$

Momentum after collision = 0
$$2x - 0.3 = 0$$
$$x = 0.15 \text{ kg}$$

(b) Case 1: Both move in opposite directions.

Momentum after collision
$$= 0.1 \times 1 + x \times (-1)$$
$$= 0.1 - x$$
So $2x - 0.3 = 0.1 - x$
$$3x = 0.4$$
$$x = 0.133 \text{ kg (3 s.f.)}$$

Case 2: Both move to the left at the same speed.

Momentum after collision
$$= x \times (-1) + 0.1 \times (-1)$$
$$= -x - 0.1$$
So $2x - 0.3 = -x - 0.1$
$$3x = 0.2$$
$$x = 0.0667 \text{ kg (3 s.f.)}$$

Case 3: Both move to the right at the same speed.

Momentum after collision
$$= x + 0.1 \times 1$$
$$= x + 0.1$$
So $2x - 0.3 = x + 0.1$
$$x = 0.4 \text{ kg}$$

Q4

(a) $2 \times 6 + 3 \times 0 = 5v$
$$12 = 5v$$
$$v = \frac{12}{5} = 2.4 \text{ m s}^{-1}$$

(b) $5 \times 2.4 = 5 \times (-0.4) + m \times 0.7$
$$12 = -2 + 0.7m$$
$$14 = 0.7m$$
$$m = \frac{14}{0.7} = 20 \text{ kg}$$

How to solve these questions

You need to describe the three ways that the particles could move.

Apply conservation of momentum.

Apply conservation of momentum.

Apply conservation of momentum.

Apply conservation of momentum.

The principle of conservation of momentum should be applied here. Note that after the collision there will be a single particle of mass 5 kg.

When using conservation of momentum here it is important to make sure that you give the velocities the correct signs.

16 PROJECTILES

Q1

(a)

$0 = 40 \sin 35° t - 4.9t^2$

$0 = t(40 \sin 35° - 4.9t)$

$t = 0$ or $40 \sin 35° - 4.9t = 0$

$t = \dfrac{40 \sin 35°}{4.9} = \textbf{4.68 seconds}$

To find the time of flight form an expression for the height of the ball and find when this is equal to zero.

(b)

Range $= 40\cos35° \times \dfrac{40\sin35°}{4.9}$

$= \textbf{153 m}$

To find the range multiply the horizontal component of the velocity by the time of flight.

(c)

$t = \dfrac{1}{2} \times \dfrac{40 \sin 35°}{4.9} = \dfrac{20 \sin 35°}{4.9}$

$= 2.341$ s

$H = 40 \sin 35° \times \dfrac{20 \sin 35°}{4.9}$

$\qquad -4.9\left(\dfrac{20 \sin 35°}{4.9}\right)^2 = \textbf{26.9 m}$

In this case the maximum height will be half way during the flight, so substitute half of the time of flight into the expression for the height.

Q2

(a)

$3 = 12\sin60° t - 4.9t^2 + 1$

$4.9t^2 - 12 \sin 60° t + 2 = 0$

$t = \dfrac{12\sin60° \pm \sqrt{(12\sin60°)^2 - 4 \times 4.9 \times 2}}{2 \times 9.8}$

$t = 1.907$ or 0.214

$t = \textbf{1.91 seconds}$

Form an expression for the height of the projectile, taking account of the initial height. Find when this is equal to 3. Use the quadratic equation formula to solve equations like this.

(b)

$x = 12\cos60° \times 1.907 = \textbf{11.4 m}$

(c)

$12 \sin60° - 9.8t = 0$

$t = \dfrac{12\sin60°}{9.8} = 1.060$ seconds

$h_{\max} = 12\sin60° \times 1.060 - 4.9 \times 1.060^2 + 1$

$= \textbf{6.51 m}$

The maximum height will be when the vertical component of the velocity is zero. Substitute this into your expression for the height.

Q3

(a)

$26\sin20° - 9.8t = 0$

$t = \dfrac{26\sin 20°}{9.8} = 0.9074$ seconds

$h_{\max} = 26\sin20° \times 0.9074 - 4.9 \times 0.9074^2$

$= \textbf{4.03 m}$

Find when the vertical component of the velocity is zero and use this to find the maximum height.

(b)

$30 = 26\cos20° t$

$t = \dfrac{30}{26\cos20°} = 1.228$

$y = 26\sin20° \times 1.228 - 4.9 \times 1.228^2$

$= \textbf{3.53 m}$

Form an equation for the horizontal motion and use this to find the time when the ball hits the wall. This time can then be used to find the height of the ball.

Q4

(a)

$0 = 1.5 - 4.9t^2$

$t = \sqrt{\dfrac{1.5}{4.9}} = \textbf{0.553 seconds}$

Note that as the initial velocity is horizontal, it does not appear in the equation for the height. Set the height equal to zero to find the time when the arrow hits the ground.

(b)

Range $= 12 \times \sqrt{\dfrac{1.5}{4.9}} = \textbf{6.64 m}$

Multiply the speed by the time of flight to find the range.

(c)

$v_x = 12$

$v_y = -9.8 \times \sqrt{\dfrac{1.5}{4.9}} = -5.422$

$v = \sqrt{12^2 + (-5.422)^2} = \textbf{13.2 m s}^{-1}$

The horizontal component of the velocity will always be 12. Calculate the vertical component using $v = -gt$. Then find the magnitude of the velocity, which gives the speed.

Q5

(a)

$0 = 8\sin\theta \times 1.2 - 4.9 \times 1.2^2$

$\sin\theta = \dfrac{7.056}{9.6} = 0.735$

$\theta = 47.3°$

Form an equation for the vertical motion, by considering when the ball hits the ground. This equation can be solved to find the angle.

(b)

Range $= 8\cos47.3° \times 1.2 = \textbf{6.51 m}$

This angle can be used with the time of flight and the speed to find the range.

17 NUMERICAL MEASURES

Answers are given to 3 s.f.

Q1

	Answers	How to solve these questions
(i)	Mean = 3.47	Find direct from calculator.
(ii)	Median is middle observation = 4	There are 15 observations so median is the 8th when they are arranged in order of magnitude.
(iii)	Mode = 4	Most frequently occurring observation.
(iv)	Standard deviation = 2.29	Direct from calculator – if in doubt use $(n-1)$ divisor.
(v)	Interquartile range = 4 − 2 = 2	Upper quartile is 12th observation in order of magnitude, i.e. 4, lower quartile is 4th observation, i.e. 2.
(vi)	Range = 9 − 0 = 9	Largest − smallest.

Q2

	Answers	How to solve these questions
(i)	Mean = 163 s	Class mid-marks are 50, 110, 170, 230 and 290. Now use calculator.
(ii)	Median = 140 + (200 − 140) × (26.5 − 22)/15 = 158 s	Either draw a cumulative frequency curve or use interpolation. There are 52 calls and so the median is halfway between the 26th and 27th. There are 22 observations less than 140 and 15 between 140 and 200.
(iii)	Modal class = '80–' s	Since all classes are of equal width choose class with highest frequency.
(iv)	Standard deviation = 69.9 s	From calculator – if in doubt use $(n-1)$ divisor.
(v)	Lower quartile = 80 + (140 − 80) × (13.25 − 5)/17 = 109.1 Upper quartile = 200 + (260 − 200) × (39.75 − 37)/9 = 218.3 Interquartile range = 109 s	Find quartiles as for median but estimating 53/4 = 13.25th and 3 × 53/4 = 39.75th observations.

Q3

Answers	How to solve these questions
Samir: mean = £1740, s.d. = £399 Marian: mean = £1640, s.d. = £238 Value of Samir's orders are, on average, higher than Marian's but they are also more variable.	From calculator. Compare means for average and standard deviations for variability.

Q4

Answers	How to solve these questions
Samir: median = £1715, interquartile range = £660 Marian: median = £1630, interquartile range = £290 Value of Samir's orders are, on average, higher than Marian's but they are also more variable.	Either draw a cumulative frequency curve or use interpolation. Answers will vary a little according to method used. Although different measures of average and variability have been used the conclusion is the same as in question 3.

Q5

	Answers	How to solve these questions
(i)	Mean = 732 hours	Class mid-marks are 674.5, 709.5, 724.5, 734.5, 749.5, 784.5. Use calculator.
(ii)	Standard deviation = 29.0 hours	
(iii)	Median = 729.5 + (739.5 − 729.5) × (50.5 − 46)/19 = 732 hours	You will get slightly different answers if you use a cumulative frequency curve.
(iv)	Lower quartile = 699.5 + (719.5 − 699.5) × (25.25 − 9)/19 = 716.6 Upper quartile = 739.5 + (759.5 − 739.5) × (75.75 − 65)/22 = 749.3 Interquartile range = 749.3 − 716.6 = 32.7 hours	
(v)	Frequency densities are 0.18, 0.95, 1.8, 1.9, 1.1, 0.26 Modal class = '730–739' hours	Since classes are of unequal width you need to find the largest frequency density.

Q6

	Answers	How to solve these questions
(a)	Median = 200 + (400 − 200) × (52.5 − 30)/31 = 345 days Lower quartile = 100 + (200 − 100) × (26.25 − 8)/22 = 182.95 Upper quartile = 400 + (800 − 400) × (78.75 − 61)/29 = 644.83 Interquartile range = 644.83 − 182.95 = 462 days	
(b)	Pitachi lasts longer, on average, before breakdown but times are more variable.	
(c) (i)	Median = 455 days Interquartile range = 798 days	Use the same method as in part (a).
(ii)	Similar on average, Pitachi less variable.	
(d)	Not possible to calculate mean and s.d. without further data on the 25 sets which had not been included.	An alternative reason is that the median and interquartile range are not affected by a few machines with very long times to first call-out.

18 PROBABILITY

Q1

(a) 0.23 — $1 - 0.64 - 0.13 = 0.23$

(b) (i) 0.0529 — 0.23^2

(ii) 0.294 — There are two ways of this occurring: $2 \times 0.23 \times 0.64$

(c) (i) 0.262 — 0.64^3

(ii) 0.295 — The probability of not paying by cheque is $1 - 0.13 = 0.87$
$3 \times 0.13 \times 0.87^2$

(iii) 0.115 — $6 \times 0.64 \times 0.13 \times 0.23$

Q2

(a) 0.512

(b) 0.256 — $0.8 \times 0.8 \times 0.2 + 0.8 \times 0.2 \times 0.4 + 0.2 \times 0.4 \times 0.8$

Q3

(a) 0.733 — $\frac{1}{6} \times \frac{2}{5} + \frac{5}{6} \times \frac{4}{5}$

(b) 0.0909 — P(windy and target) = P(target) × P(windy|target)
$\frac{1}{6} \times \frac{2}{5} = 0.7333 \times$ P(windy|target)

Q4

(a) 0.008 — 0.2^3

(b) 0.142 — $0.2^3 + 0.35^3 + 0.45^3$

(c) 0.0563 — P(all water and all same) = P(all same) × P(all water all same)
P(all water and all same) = P(all water) = 0.008
$0.008 = 0.142 \times$ P(all water|all same)

Q5

(a) 0.667 — 80 out of 120 recruits were offered permanent employment.

(b) 0.392 — 47 recruits were < 20 and were offered permanent employment.

(c) 0.967 — All but $3 + 1 = 4$ recruits were either < 20 or were not offered a further trial (or both).

(d) 0.733 — There were 45 recruits aged 20 or over (Q'), of whom 33 were offered permanent employment.

(e) 0.25 — $P(F) = \frac{45}{120}$ P(F∩R) = P(F) × P(R), F and R are independent.

(f) 0.3 — P(F∩Q) = P(F) × P(F | Q)

(g) 0.133 — P(S∩F) = P(F) × P(S | F)

Q6

(a) (i) 0.01475 — $0.01 \times 0.98 + 0.99 \times 0.005$

(ii) 0.664 — P(D∩+ve) = 0.01×0.98 = P(+ve) × P(D | +ve)

(b) 0.653 — $0.6644 \times 0.98 + (1 - 0.6644) \times 0.005$

19 BINOMIAL DISTRIBUTION

Q1

(a) 0.733 — B(20,0.09) You can read the answer from tables.
P(4 or fewer) − P(3 or fewer)

(b) 0.0703

Q2

(a) 0.253 — B(8,0.43) tables won't include this, so you will have to use the formula.

(b) 0.0784 — P(0) + P(1)

(c) 0.986 — $1 - $ P(7) − P(8)

Q3

(a) 0.850 — B(10,0.3), which you can read from tables.
P(4 or fewer) − P(3 or fewer)

(b) 0.200 — P(5 or fewer) − P(1 or fewer)

(c) 0.303 — B(10,0.7) but this is not in tables.

(d) 0.0473 — 4 or fewer 'failures' → 6 or more 'successes' → 1 − 5 or fewer 'successes'.

6, 2.05 — $np, \sqrt{np(1-p)}$

Q4

(a) (i) 0.210 — B(16,0.15): $1 - $ P(3 or fewer), but not all binomial tables will include this.
P(4 or fewer) − P(3 or fewer)

(ii) 0.131

(b) (i) 0.934 — B(5,0.2101) : you will have to calculate this. Use as many significant figures as your calculator will give you for p.
5×0.2101 (The expected number is the mean.)

(ii) 1.05

Q5

(a) 1.525, 1.48 — From the calculator; the standard deviation with divisor n giving 1.47 is also acceptable.

(b) 0.381 — 61 out of 160 jumps were disallowed.

(c) 1.525, 0.944 — $np, np(1-p)$

(d) No, the standard deviation predicted by binomial $\sqrt{0.944}$ = 0.971 is a long way from the observed value of 1.48.

(e) Some competitors are more likely to have jumps disallowed than others → p is not constant.

Q6

(a) (i) 0.0905 — B(25,0.3) which you can read from tables.
19 not agree → 6 agree, P(6 or fewer) − P(5 or fewer)

(ii) 0.147

(b) (i) Yes, n fixed, p constant

(ii) No, n not fixed

Answers

20 Normal distribution

How to solve these questions

Q1
$z = -1.6$
P$(X < 32) = 0.0548$ (3 s.f.)

$P(X < 32) = 1 - 0.945\,20 = 0.0548$

Q2
$z_1 = -1.5$, $z_2 = 2.0$
P(component functions satifactorily) = **0.910**

P(component functions satisfactorily)
$= 0.977\,25 - (1 - 0.933\,19) = 0.910$

Q3
(a) $z = -0.5454$
The proportion with a measurement below 250 is
$1 - 0.707 = 0.293$

An answer based on rounding z to 0.55 would have been accepted.
Interpolating between $z = -0.54$ and $z = -0.55$.

(b) The measurement exceeded by 20% of smokers is $310 + 0.8416 \times 110 = 403$

Q4
(a) $z = 1.5$
probability $< 15800 = 0.933$

(b) $z_1 = -1.75$, $z_2 = 1.5$
P(lifetime between 14500 and 15800 hours) = **0.893**

P(lifetime between 14500 and 15800 hours)
$= 0.933\,19 - (1 - 0.959\,94) = 0.893$

(c) For the 84th percentile, $z = 0.9945$
The lifetime is **15600 hours** (3 s.f.).

Remember that the 84th percentile is the value which exceeds 84% of the population.
The lifetime $= 15200 + 0.9945 \times 400 = 15597.8$ hours.
Round to three significant figures.

Q5
(a) (i) $z = 1.25$
P(time spent in the pool will be < 95)
$= 0.894$

(ii) $z_1 = -0.25$, $z_2 = 1.25$
P(spending between 65 and 95 minutes) = **0.493**

P(spending between 65 and 95 minutes)
$= 0.894\,35 - (1 - 0.598\,71) = 0.493$

(b) The maximum time a user can spend in the pool is 60 minutes. A model with mean 70 minutes could not apply.

(c) 99.9% of normal distribution exceeds $\mu + 3\sigma$. The model would be plausible if a user has at least $70 + 3 \times 20 = 130$ minutes in the pool. **The user must enter by 6.50 p.m.**

The maximum time a user can spend in the pool is 60 minutes. A model with mean 70 minutes could not apply.

A less cautious answer based on 95% or even 90% of the distribution would be acceptable.

Answers

How to solve these questions

Q6
(a) (i) $z = 1.4$

The proportion requiring large jackets is $1 - 0.919\,24 = 0.0808$

(ii) $z_1 = -0.6$, $z_2 = 1.4$
The proportion of medium jackets is **0.645**.

The proportion of medium jackets is $0.919\,24 - (1 - 0.72575) = 0.645$.

(b) $z = 2.0537$
The minimum chest measurement for extra large is **112.3 cm**.

The minimum chest measurement for extra large is $102 + 5 \times 2.0537 = 112.3$ cm.

(c) The proportion of people needing small jackets is $1 - 0.72575 = 0.27425$, the median of these must exceed $0.27425/2 = 0.137125$ of the population.

Median measurement of small jackets is $102 - 1.09 \times 5 = 96.5$ cm.

Using interpolation $z = -1.09$.
Rounding the probability to 0.14 and obtaining $z = -1.08$ is also acceptable.

21 Correlation and regression

Q1
0.868
Heavier ewes tend to have heavier lambs.

You can use your calculator to find this answer.
Positive value close to 1 shows strong association.

Q2
(a)

(b) $y = 1.09x + 0.246$

From calculator

(c) Miriam's estimates lie close to the regression line. This line is a little above the ideal line $y = x$. **She tends to overestimate slightly.**

You can see this from your diagram.

Q3

(a)

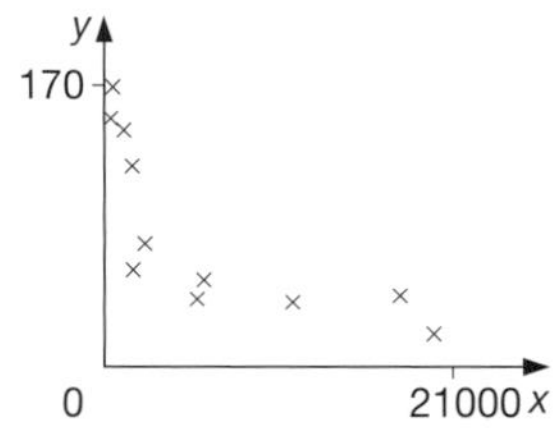

(b) For countries with low GDP per capita a small increase in GDP is associated with a large reduction in mortality before age 5. This effect is less marked for countries with a higher GDP per capita.

(b) (i) −0.739

(ii) The relationship appears not to be linear so the product moment correlation coefficient is not appropriate.

(c) −0.936

(d) Both show an association between low GDP and high mortality. Spearman's rank correlation coefficient is much closer to −1 because although the relationship is not linear, the ranking of the countries on GDP is almost the reverse of the ranking on mortality.

How to solve these questions

The non-linear relationship which is clear from the scatter diagram is the key to this question.

From calculator

First rank both sets of data. There are no ties so your calculator will give you the same answer as using the formula based on Σd^2.

Q4

(a)

$$b = \frac{484 - \dfrac{143 \times 391}{15}}{2413 - \dfrac{143^2}{15}} = -3.089\,86$$

$$a = \frac{391}{15} + \frac{3.089\,86 \times 143}{15}$$

$$= 55.5234$$

$$y = 55.5234 - 3.089\,86x$$

$$h - 100$$
$$= 55.5234 - 3.089\,86(s - 20)$$

$$h = 217 - 3.09s$$

(b) The slope of the line is an estimate of the **reduction in life (hours)** for each revolution per minute increase in speed.

(c) **125 hours**

Q5

(a) *See diagram opposite.*

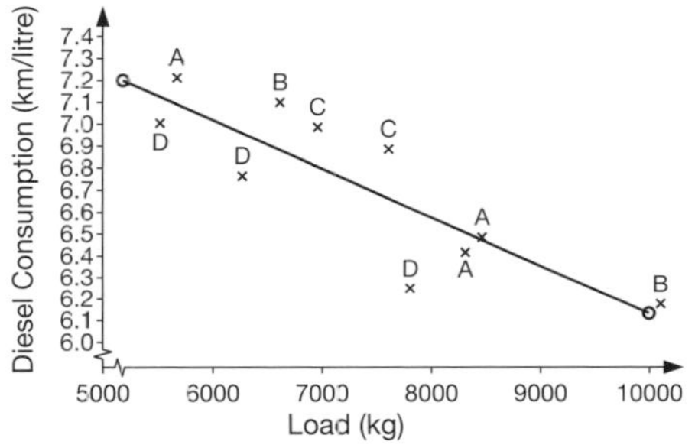

(b) $y = 8.36 - 0.000\,222x$

(c) The intercept estimates the diesel consumption of a lorry with no load. The slope estimates change in diesel consumption for each additional kg carried.

(d) 30 000 kg is outside the range of the observed data.

(e) Dan's diesel consumption is always less than predicted by the regression line.

(f) The load may affect diesel consumption but diesel consumption cannot affect the load.

How to solve these questions

Unfortunately the raw data has not been given so you will have to substitute in the formulae instead of using your calculator directly.

Keep as many figures as possible for these calculations.

Now substitute for x and y to obtain the equation in terms of h and s as required.

This is the required equation so now round your values to three significant figures.

Substitute $s = 30$ in the equation.

Since diesel consumption is measured in km/litre this is undesirable.

22 Algorithms

Q1

63	32	70	26	59	41	17
32	63	70	26	59	41	17
32	26	63	70	59	41	17
32	26	59	63	70	41	17
32	26	59	41	63	70	17
32	26	59	41	17	70	63
32	26	59	41	17	63	70

32	26	59	41	17
26	32	59	41	17
26	17	32	59	41

26	17	32	59	41
17	26	32	41	59

How to solve these questions

The quicksort algorithm takes the first number as a pivot and separates the list into two subsets; those numbers larger than the pivot and those smaller than the pivot. Show the pivot element – here it is 63.

We now have two lists 32, 26, 59, 41, 17 and 70

Repeat the steps for the two lists. We only need to sort the first list using the pivot 32.

We now have the subsets 26, 17 and 59, 41.

Use the pivots 26 and 59 for the final steps in the algorithm.

Q2

A=1	Yes	11	No	21	No	31	No	41	No	51	No
2	Yes	12	Yes	22	No	32	No	42	No	52	No
3	Yes	13	No	23	No	33	No	43	No	53	No
4	Yes	14	No	24	No	34	No	44	No	54	No
5	Yes	15	Yes	25	No	35	No	45	No	55	No
6	Yes	16	No	26	No	36	No	46	No	56	No
7	No	17	No	27	No	37	No	47	No	57	No
8	No	18	No	28	No	38	No	48	No	58	No
9	No	19	No	29	No	39	No	49	No	59	No
10	Yes	20	Yes	30	Yes	40	No	50	No	60	Yes

The algorithm produces the factors of 60.

How to solve these questions

Follow through the algorithm carefully noting the outcome of each integer A. When you have completed the algorithm look at the pattern of numbers to deduce the outcome of the algorithm.

Answers

Q3

How to solve these questions

For problems of this type, look for a Hamiltonian cycle which then forms the basis of the re-drawing to show that the graph is planar.

Remember Euler's formula $v - e + f = 2$ is a formula connecting the number of vertices (nodes), edges (arcs) and faces for a planar graph.

(a)

The graph is planar.

$v = 5, e = 9, f = 6, v - e + f = 2$

(b)

The graph is planar.

$v = 6, e = 12, f = 8, v - e + f = 2$

Q4

(i) 11 names so the middle name is $\frac{1}{2}(1 + 11)$

= 6 JOYCE

Reject ALLAN to JOYCE since SUSIE comes after JOYCE

  7. HO-YI
  8. REBECCA
  9. SARAH
  10. SUSIE
  11. WILLIAM

the middle name is $\frac{1}{2}(7 + 11) = 9$ SARAH

Reject HO-YI to SARAH as SUSIE comes after SARAH

  10. SUSIE
  11. WILLIAM

the middle name is $\frac{1}{2}(10 + 11) = 11$ WILLIAM

Reject WILLIAM. SUSIE is found.

(ii) 10000, 5000, 2500, 1250, 625, 312, 156, 78, 39, 19, 9, 4, 2, 1   14 iterations.

How to solve these questions

The list is first ordered in ascending or descending order. The binary search algorithm finds the middle item in a list (or the one next to the middle item). If this is not the correct item then we reject the half of the list that does not contain the item. You continue halving the remaining list using the middle item until the correct item is found.

The list reduces by approximately one half on each iteration.

Q5

(a)

90	35	105	90	120	45	60	100
90	105	35	90	120	45	60	100
90	105	90	35	120	45	60	100
90	105	90	120	35	45	60	100
90	105	90	120	45	35	60	100
90	105	90	120	45	60	35	100
90	105	90	120	45	60	100	35

90	105	90	120	45	60	100	35
105	90	90	120	45	60	100	35
105	90	120	90	45	60	100	35
105	90	120	90	60	45	100	35
105	90	120	90	60	100	45	35

105	90	120	90	60	100	45	35
105	120	90	90	60	100	45	35
105	120	90	90	100	60	45	35

105	120	90	90	100	60	45	35
120	105	90	90	100	60	45	35
120	105	90	100	90	60	45	35

120	105	90	100	90	60	45	35
120	105	100	90	90	60	45	35

120	105	100	90	90	60	45	35
120	105	100	90	90	60	45	35

(b) Full-bin algorithms

Tape 1 120 60
Tape 2 90 90
Tape 3 100 45 35
Tape 4 105

First-fit decreasing algorithm

120, 105, 100, 90, 90, 60, 45, 35

Tape 1 120 60
Tape 2 105 45
Tape 3 100 35
Tape 4 90 90

How to solve these questions

For a bubble sort one number is placed in its correct position in each pass through the algorithm. The algorithm starts by comparing the first two numbers in the list; if the second is larger than the first then the numbers are exchanged. It then compares the second and third and exchanges them if necessary, and so on through the list. In the question here the number 35 finds its way to the bottom.

Now we repeat the steps in the algorithm to see that 45 goes to its correct position. And so on.

For the first fit decreasing bin packing algorithm we first reorder the list of numbers into descending order. Now we work through the list placing an item in a bin that has room for it.

---

23 NETWORKS

Q1

(a) Each edge (arc) contributes 2 to the sum of degrees, hence this sum is even. Thus there must be an even (or zero) number of vertices of odd degree.

(b) Since B and C are vertices of odd degree we enter and leave them twice.

Repeat BC if $BA + AC > BC$ i.e. $x > 9$

Repeat BC: $x > 9$: length = sum of edges + BC
$= 0.5x - 26 = 100$
$x = 12$

Repeat BA and AC: $x < 9$; length $= 11.5x - 35 = 100$
$x = 11.7$ reject this solution as inconsistent.

Thus $x = 12$.

Q2

(a) In Prim's algorithm the spanning tree grows in a connected fashion adding a new edge to the tree whereas for Kruskal's algorithm the shortest arc is added whether it connects or not.

There is no need to check for cycles when using Prim's algorithm.

Prim's algorithm starts with a vertex whereas Kruskal's algorithm starts with an edge.

Prim's lends itself to a matrix representation.

(b) (i)

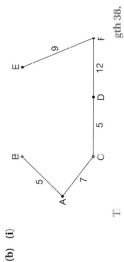

length 38.

(ii) AB – 5, CD – 5, AC – 7, EF – 9; BC – 11, DF – 12, BE – 13, DE – 15, CE – 18

Spanning tree is AB, CD, AC, EF, DF (not BC because a cycle is formed).

How to solve these questions

(a) Identify any vertices (nodes) with odd degree – the total length of the route inspection problem will then be the sum of the weights of the edges plus repeated edges. You need to identify possible repeats of edges. In this problem we could either repeat edge BC or the two edges AB and AC. Remember that we need to return to the starting vertex. The solution repeating BA and AC is inconsistent because x cannot be equal to 11.7 and less than 9.

Choose any vertex to start with, in this solution we have chosen E. Join the shortest edge to E, i.e. EF. Now join the shortest edge to E or F which is not used, i.e. FD. Now join the shortest edge to E, F or D that is not used, i.e. DC. Now join the shortest edge to E, F, D or C that is not used, i.e. CA. Now join the shortest edge to E, F, D, C or A that is not used, i.e. AB. The solution is now complete as all the vertices are connected.

For Kruskal's algorithm we first list the edges in order. Now form the spanning tree by selecting the edges in turn: AB, CD, AC, EF, DF (not BC because a cycle is formed). We get the same spanning tree as for Prim.

Draw the precedence network to show the necessary steps in a project. Label the network to show the earliest start time for each activity. This is the forward pass.

Now work backwards from the finish time labelling each vertex with the latest finishing time.

The critical path is the minimum time needed to complete the project and includes all the essential activities. These must be started on time to avoid delaying the whole project.

The maximum time by which an activity can be delayed is the float. In this question:
float for C = latest start time for F – earliest start time for C – time for C

Show the earliest possible start time for each person.

Use the float for each non-critical activity to adjust the start and finish time for each worker. In this case the second worker can be used to complete tasks BCFH.

Answers

24 CRITICAL PATH ANALYSIS

Q1

(a)

F(1), G(6), 10|10, C(4), 8|9, A(4), 4|4, D(3), B(2), 0|0, E(1), 2|3, 4|4

(b) A, G

(c) $9 – 4 – 4 = 1$

Q2

(a) Critical activities ADEG
Length of critical path 17 hours

(b) Floats:
B: $7 – 0 – 6 = 1$
C: $11 – 0 – 5 = 6$
F: $13 – 5 – 2 = 6$
H: $17 – 5 – 3 = 9$

(c) 0 1 2 3 4 5 6 7 8 9 10 11 12 13 14 15 16 17 18 19
A, D, E, G, B, C, F, H

(d) 0 1 2 3 4 5 6 7 8 9 10 11 12 13 14 15 16 17 18 19
A, D, E, G, B, C, F, H

Only two workers are required to complete the project in the given time.

The labelling procedure uses three boxes on each vertex,

This side gives the order of labelling
This side shows the distance from the starting vertex
This box is for working

Start by giving a permanent label 0 to the starting vertex A. In the working space give temporary labels to the vertices connected to A. Give a permanent label to the smallest working value, i.e. to C. Now give temporary labels to the vertices connected to C. Assign a permanent label to the vertex with the smallest temporary label, i.e. B. Continue with the labelling until each vertex has a permanent label.

The shortest path is found by working backwards matching the edge weights to the differences between the permanent labels.

The first step is to look for a feasible flow through the network. In this solution we have chosen a flow of 3 along SACT. The forward arrows show any remaining capacity (excess capacity) whereas the backward arrows show the flow along the arc. An arc with zero excess capacity is said to be saturated. Now improve the flow by looking for paths from S to T which consist entirely of unsaturated arcs – these are called flow-augmenting paths. Use them to increase the flow through the network. When no more flow-augmenting paths can be found the flow is a maximum.

Answers

Q3

3|7, 7, 5|13, 13, F 6|16, 4|12, B, D, 8, 3, 5, A 1|0, 7, 5, 9, C 2|4, 4, E

The shortest path ACDF has length 16.

Q4

(a) (network diagram S, A, B, C, D, T)

The maximum flow is 10.

(b) The cut through AC, BC and BD has value 10 – max flow/min cut theorem proves the flow is a maximum.

The max flow/min cut theorem can be used to confirm the result.

25 OPTIMISATION

Q1

(100)A
(800)B
(100H)C
(400)D
(*)E
(*)F

100(F)
200(C)
400(E)
800(E)
1500(D)
100H(F)

The initial matching M chosen is A–100, B–800, C–100H, D–400.

The alternating paths are 200–C–100H–F and 1500–D–400–E.

Edges of M that are not in the alternating paths are: A–100 and B–800.

Edges in the alternating path but not in M are: C–200, F–100H, D–1500 and E–400

A complete solution is:
A–100 m
B–800 m
C–200 m
D–1500 m
E–400 m
F–100 m hurdles

Q2 (a)

	Component A	Component B	Profit
Analogue (x)	2	3	£20
Digital (y)	4	1	£25
Totals	50	24	T

$2x + 4y \leq 50$
$3x + y \leq 24$
$x + y \leq 20$
$x \geq 2, y \geq 2$
$T = 20x + 25y$

How to solve these questions

Draw the bipartite graph using the information in the question and choose an initial matching M. The left vertices are called set A and the right vertices are called set B.

The labelling procedure starts by labelling with (*) all the vertices in A that are not in M.

Step 2: choose a newly labelled vertex in A, say a, and label with (a) all the unlabelled vertices in B joined to a by an edge not in M. Repeat for all newly-labelled vertices in A before going on to the next step.

Step 3: choose a newly labelled vertex in B, say b, and label with (b) all the unlabelled vertices in A joined to a by an edge in M. Repeat for all newly-labelled vertices in A before going on to the next step.

Step 4: repeat steps 2 and 3 until no further labelling is possible.

Now find an alternating path starting from a vertex in B that was not in the initial matching M. Follow the edges back to (*) using the labels.

An improved matching consists of:
- the edges of M which are not in the alternating path;
- the edges in the alternating path but not in M.

In these types of problems there may be more than one solution.

It is often helpful to draw up a table showing the information given in the question.

Formulate the objective function and the constraints.

In this problem x and y are both greater than or equal to 2 (often x and y need to be positive).

(b)

(c) **Minimum value of T is £90 at (2, 2).**
The vertex for a maximum is (4.6, 10.2) which is not possible.
At (3, 11) $T = 335$; at (4, 10) $T = 330$ and at (5, 9) $T = 325$.
The maximum value for T is £335.

Q3

$3x + 4y + u = 48$
$2x + y + v = 17$
$3x + y - w = 24$
$f - 9x - 4y = 0$

b.v.	x	y	u	v	w	value	
u	3	4	1	0	0	48	$48 \div 3$
v	2	1	0	1	0	17	$17 \div 2$
w	3	1	0	0	1	24	$24 \div 3$
f	−9	−4	0	0	0	0	

b.v.	x	y	u	v	w	value	
u	3	4	1	0	0	48	
v	2	1	0	1	0	17	
w	3	1	0	0	1	24	$R3 \div 3$
f	−9	−4	0	0	0	0	

table continues

How to solve these questions

Draw a graph showing the constraints as straight lines, e.g. $2x + 4y = 50$. For each line replaced by $2x + 4y = 50$ is shade the region that does not apply – then you can show the feasible region.

Draw the objective function line.

To find the optimal value of T consider the extreme points of the feasible region.

In this case the optimal values of x and y for maximum T are not integers. It is difficult to make 4.6 analogue televisions! You need to investigate the integer coordinate points close to (4.6, 10.2).

Introduce slack variables u, v and w to give the system of linear equations.

Draw the simplex tableau.
Choose to increase either x or y. Since increasing x will have a larger increase in f than increasing y we choose to increase x.

Pivot about row 3 since the entry in the final column for row 3 divided by the entry in the x column is the least value.

b.v.	x	y	u	v	w	value	
u	-3	4	1	0	0	48	R1 – 3R3
v	-2	1	0	1	0	17	R2 – 2R3
x	1	$\frac{1}{3}$	0	0	$\frac{1}{3}$	8	
f	-9	-4	0	0	0	0	R4 + 9R3

Pivot about row 2.

b.v.	x	y	u	v	w	value	
u	0	3	1	0	-1	24	24 ÷ 3
v	0	$\frac{1}{3}$	0	1	$-\frac{2}{3}$	1	$1 \div \frac{1}{3}$
x	1	$\frac{1}{3}$	0	0	$-\frac{1}{3}$	8	$8 \div \frac{1}{3}$
f	0	-1	0	0	3	72	

b.v.	x	y	u	v	w	value	
u	0	3	1	0	-1	24	
v	0	$\frac{1}{3}$	0	1	$-\frac{2}{3}$	1	$R2 \div \frac{1}{3}$
x	1	$\frac{1}{3}$	0	0	$-\frac{1}{3}$	8	
f	0	-1	0	0	3	72	

b.v.	x	y	u	v	w	value	
u	0	3	1	0	-1	24	R1 – 3R2
v	0	1	0	3	-2	3	
x	1	$\frac{1}{3}$	0	0	$\frac{1}{3}$	8	$R3 - \frac{1}{3}R2$
f	0	-1	0	0	3	72	R4 + R2

Answers

b.v.	x	y	u	v	w	value
u	0	0	1	-9	5	15
y	0	1	0	3	-2	3
x	1	0	0	-1	1	7
f	0	0	0	3	1	75

The maximum value for f is 75 and occurs when $x = 7$, $y = 3$, $u = 0$, $v = 0$ and $w = 0$.

How to solve these questions

Since each entry in the final row is positive we have achieved the maximum value for f. If we increase v or w then we decrease f, since the bottom row gives

$$f = 75 - 3v - w$$

WITHDRAWN

THE LIBRARY
SAINT FRANCIS XAVIER
SIXTH FORM COLLEGE
MALWOOD ROAD, SW12 8EN